DETERMINATION OF PINUS SPECIES BY POLLEN STATISTICS

DETERMINATION OF PINUS SPECIES BY POLLEN STATISTICS

BY
WILLIAM S. TING

UNIVERSITY OF CALIFORNIA PRESS
BERKELEY AND LOS ANGELES
1966

University of California Publications in Geological Sciences

Volume 58

Approved for publication March 26, 1965
Issued October 28, 1966

Price, $6.00

University of California Press
Berkeley and Los Angeles
California

Cambridge University Press
London, England

PRINTED IN THE UNITED STATES OF AMERICA

CONTENTS

Abstract 1
Introduction 1
Acknowledgments 4

PART I

Pollen morphology 6
The distal membrane of corpus 6
Pattern of the cappa 7
Crests and sacci attachment 8
Abnormal pollen grains 9
Shape and orientation of the pollen grain 10
Other features 10
Techniques 12
Reference pollen material 13
Physical and chemical change in size of pollen grains 13
Changes owing to moisture, heat, and acidity 13
Elasticity of pollen wall 15
Comparison of sizes of fossil and modern pollen grains 15
Some investigations regarding acetolysis 17
The use of hydrofluoric acid and heavy liquids 18
Mounting media and size changes of pollen grains in glycerine-jelly mounts 19

PART II

Statistics 22
Measurements 22
The scattergram and sample size 22
The class interval 27
Statistical data 27
Ratio index 30
Cos A index 31
Spectrum and modal distribution 33
Coefficient of variance 37

PART III

Specific identification of single species 48
Introduction 48
Example I 48
Statistics 48
Morphology 49
Identification 49
Comments 49
Example II 51
Morphology 51
Statistics 51
Identification 51

Example III 51
Example IV 53
Morphology 53
Statistics 53
Example V 54
Sample 1. *Pinus murrayana* 54
Sample 2. *Pinus flexilis* 54
Sample 3. *Pinus jeffreyi* 54
Sample 4. *Pinus coulteri* 56
Sample 5. *Pinus monophylla* 56

PART IV

Specific identification of mixed samples 57
Introduction 57
Class-C index 57
Estimation of the number of components 61
Mathematical considerations 62
An illustration 63
Hypotheses 1 and 7 64
Application to a mixed sample of known *Pinus* 67
Fitting of B 69
Fitting of A 71
Reestimation of y and estimation of z 72
Calculations and interpolations 73
Fitting by coarse grouping 74
Conclusions 74
Identification of the unknown 76
Preliminary considerations 76
Modal distribution 77
Analysis of spectrum II 77
Specific identities and quantities 79
Analysis of the spectrum of saccus measurements 80
Number of components 81
Ratio index and cos A index 82
Example I. Fossil sample 1, Merced 2163 85
Preliminary analysis of the spectra 85
Fitting of spectrum $A + B$: first trial 89
Fitting of spectrum $A + B$: second trial 93
Fitting of spectrum $A + B$: third trial 97
Fitting of spectrum C 99
Conclusions 99
Example II. Fossil sample 2, Merced 3022 101
Initial grouping and fitting A and C spectra 101
Components of spectrum $B4$ 102
Components and their quantities 104
Other considerations 105

Summary 105

APPENDIXES

I. Statistics of individual reference samples of modern *Pinus* pollen grains 107

II. Frequency spectra of breadth of corpus from pollen grains of samples of *Pinus* species 116

III. Posterior probabilities and Mahalanobis distance from group of the pollen groups and species of *Pinus* 128

IV. The degree of resemblance measured by the gamma index 133

V. Comparison of sizes of fresh and acetolyzed pollen grains of *Pinus* 139

VI. Discussion on papers by B. B. Christensen (1946) and S. T. Anderson (1960) concerning size of pollen grains 142

VII. Comparison of pollen counts of samples treated by potassium hydroxide and by acetolysis 151

VIII. Fossil pollen size changes under various treatments 155

IX. The density separation of pollen grains 158

Symbols and abbreviations 161

Glossary 163

References 165

Plates 169

DETERMINATION OF PINUS SPECIES BY POLLEN STATISTICS

BY

WILLIAM S. TING

ABSTRACT

A METHOD FOR determining species of California pines from statistics of their pollen is devised and discussed in detail. Examples of the procedure are given to illustrate the application of the method to unknowns containing only one species, and to mixed samples containing more than three species. The method has been tested, and the results are encouraging since species identification of pines by pollen has not been considered possible by most palynologists.

In this study, more than 300 samples of 19 species, including varieties and hybrids of pines mainly of California, were collected and studied. About 280 samples were used as references. A comprehensive collection of reference material is mandatory in order to determine the variation of the species over its range.

The chosen technique of preparation was acetolysis and glycerine-jelly mounts. Other techniques and mounting media may be used if a rigid standard is maintained throughout.

Three measurements of the grain are used: breadth of corpus, breadth of saccus, and overall breadth. While the dimensions and statistics of the corpus constitute the major criteria, data of saccus or of overall breadth may also be used. In order to overcome the lack of fine discrimination of a univariate analysis, a new index—the cos A index—was devised and found to be highly discriminating for specific identification. This index is based on three ratios of the three measurements, and is in essence a sort of discriminating function.

In application to fossil assemblages that are undoubtedly mixed, the coefficient of variance is found to be an indicator of mixing. Solution of mixed assemblages by iteration shows that the method can resolve a composite frequency distribution into its componential parts, which give the identity of the taxa and also the quantity of each.

INTRODUCTION

Commencing in 1958, I began to study spore-pollen floras from Late Pliocene and Early Pleistocene sediments of California which contain a *Pinus-Abies* association (Axelrod and Ting, 1960, 1961). It was immediately apparent that unless a method could be devised which would enable one to identify species, the problem of interpreting past vegetation and climate could not be solved, because in California different species of these genera range from sea level to timberline, occupying distinct vegetation and climatic zones. As no method of specific identification had yet been discovered, it was decided that such an attempt should be made even though the characters of pine pollen appear to be very uniform. My first attempt was a simple semistatistical method founded solely on (*a*) the mean breadth of corpus, and of overall size of grain, and (*b*) a scattergram termed "Distributional Diagram" (in Axelrod and Ting, 1960). Although the method was very simple, and was based on a woefully meager amount of reference material, the results were encouraging. This became apparent because the floras east of the Sierra Nevada, distributed 150 miles north–south and 50 miles east–west, show gradual ecological gradients in terms of angiosperm communities, and the gradients were wholly consistent with the species of pines which were identified. Furthermore, since at that time I had no knowledge of the zonal relations of California vegetation, the identification of species of *Pinus* (and *Abies*) could scarcely be ascribed to ecologic bias on my part. Following those reports, the study was extended to cover some 19 species of pine as well as other genera of conifers in California and bordering areas. This continuing study has resulted in the development of the method described here.

The reason why a statistical approach has been chosen should be explained. There have been reports, for instance by Hansen (1947) and Erdtman (1931), that certain species can be distinguished by means of their individual morphological features. To a certain extent this is true, but these examples were based on only a limited number of unknowns, or with known sampling. I examined different individual morphological features of the pollen grains in *Pinus* and found that only one or two of the features are relatively constant and can be used only for subgeneric groupings. Other characters may be useful in distinguishing pollen grains of a particular species from another, but are not universally applicable. Still other characters are present in only some grains of a species and are therefore inadequate as criteria for identification.

The problem of species identification can, of course, be approached in other ways. Pollen chemistry in the sense of the response to staining, or electron microscopy, can be of value. The difficulty with these methods lies partly in the time, labor, and cost involved in building up a substantial reference collection, in the availability to the average palynologist of expensive equipment, and also in the necessity of preserving the fossil specimens. The wish, therefore, has been to devise a simple method of identification which can be accomplished with minimum effort, cost, and time, and one that is easily available to all investigators. No claim is made here that the present method has been wholly refined. Rather, there is immediate need of testing the statistical tenets on which it is based so that it can be refined further.

Determination of pine pollen by statistical methods has been attempted by investigators in this country (Cain, 1940; Hansen, 1947; Grayson, 1954; Whitehead, 1964).* A search of the literature will quickly demonstrate that the method has also been explored by palynologists of many other countries (e.g., Hormann, 1929; Bertsch, 1931; Jaeschke, 1935; von Sarntheim, 1936; Christensen, 1946; Buell, 1946; Federova, 1952; Pons, 1956; Harris, 1956; Martin, 1959; Aytug, 1960). Statistics based on measurements are easy to gather. In the literature, measurements of pines have been made on length of corpus, height of corpus, width of corpus, width of saccus, mean value of any sample of the above features, angle between corpus and saccus, and overall length or breadth—all used individually or in combination. The recorded measurements vary greatly because of different methods in chemical treatment, in mounting, in lapsed time between mounting and measuring, in thickness of the mount, in chemicals used for storage

* Whitehead's paper, in which he discusses the problems encountered in using size frequency as the sole means of identification of fossil *Pinus* pollen, was published just as this work was submitted for publication. Much of his discussion is based upon the results of statistics of only one dimension. He too is concerned with the possible physical-chemical change of pollen grains which took place during fossilization, subsequent laboratory treatments, and other procedures. I discuss the shortcomings of using one dimension and frequency curves as the only means of identification and also discuss in full the aspects of physical-chemical changes in pollen grains in the sections on technique (Part I) and statistics (Part II) and in Appendixes III–VIII.

of prepared pollen, in the optical view measured, and even in the scale used for measuring. Some of the divergent results also relate to whether the original reference material was from herbarium sheets or fresh from trees, whether the plant was correctly identified initially, and whether the amount of reference material of a given species was sufficient to include all the important racial variations of the taxon. The maturity of herbarium specimens is, of course, of concern and as far as my collecting goes, immature specimens have not been included. The assumption is herein made that if a standard procedure is strictly adhered to, statistics should give a more consistent measure than morphological traits within the limits of confidence controlled by probability.

Before the summer of 1962, when the present method was finally emerging from the statistical analysis, I received a series of pollen samples to be used as tests. One of these tests covered several genera and species other than *Pinus,* together with four species of pine. Most of them were identified correctly, but three of them—two pines and a spruce—were in error in terms of specific identification. The size of pollen grains of those mistakenly identified displayed variation well outside that represented in the original reference material. This discovery initiated the building up of the reference slides from 20 to 280 samples of the California species of *Pinus,* and led to the conclusion that there are many variant forms (or races) of California pines. It is now clear that only by wide sampling of all possible races of pine can the total characteristics of the species be properly understood. Without this knowledge, specific identification by the method described here is not always possible.

For my own use I arranged the data in a number of tables and keys for easy reference. These are not included here. Instead, two tables listing all the data are included (appendix tables 1, 2). These will be useful to others working on problems connected with the California pines.

Based on the larger sampling on hand, I conducted some additional tests in 1962. The first test consisted of six unknown pollen samples from six pines, only one of which was a California species, and it was correctly identified. The other five were European and Asian species, not covered by reference, and were therefore not correctly identified. Another test consisted of ten unknown samples of pine, nine of which were correctly identified. In the case of the tenth, there was doubt about the identity of the original specimen from which the pollen was collected. Unknowns I prepared myself from fresh pollen supplied by the Institute of Forest Genetics were all correctly identified except one, which so far has been incorrectly identified twice. This is *Pinus contorta,* mistaken for *P. murrayana.* The latest test, done in 1963, consisted of five species and all were correctly identified. The history of success is not without its setbacks. In another series of slides made for testing, all were incorrectly identified. This was because of faulty mounting of the slides, but the incident served to show that identification is a delicate matter. Slight irregularity in processing may offset the measurement data a great deal. Reflecting on all these tests, it is obvious that the results were not accidental; 40 out of 45 have been correctly identified.

The method is now more or less in its final form. It can be adapted for studies of pollen grains of *Pinus* in other areas. A good statistical reference collection for the species, deemed relevant, is necessary. The technique can be changed to accommodate individual preference providing the references and the unknown are treated by one standard technique.

The identification of a modern species by its pollen shows that statistics may serve as a taxonomic discriminant in pines, and two computational programs (Appendixes III and IV) confirmed this. The application of the method to identification of pine species in modern pollen grain studies can now be attempted with some confidence.

There is much still to be done on the identification of fossil pollen grains. The method of iteration, though statistically similar to simulation, has been misunderstood as "cutting out designs." It is, as one reviewer remarked, designing probable subspectra from the basic fossil assemblage spectrum and trying to account for every specimen in the tally with as few species as possible. One will find that this is exactly true, for all we have as a basis is the spectrum of fossil pollen grains, and all we can do is simulate the conditions under which such a spectrum can be duplicated by a certain number of known spectra from established references. I include the symbols and abbreviations used in statistics for the benefit of the reader unfamiliar with the terminology and to clarify the specific usage I make of some of the terms. The palynological terms in the text are from Erdtman (1957), and may also be found in Kremp (1965) and Gray (1965). The chemical terms are the ones in common use in palynology and are not included. They may be found in Gray (1965).

Acknowledgments

I am most indebted and grateful to my wife, Katherine S. Bao, M.D., and to Professor Daniel I. Axelrod for their constant encouragement and inspiration. They provided the stimuli and drive without which I might have been discouraged by initial failures.

I wish especially to thank Professor G. Ledyard Stebbins, University of California, Davis, for arranging two tests that took up a great deal of his time in gathering material, checking the identity of the specimens, and making available good and unbiased material for testing. To Professor G. Erdtman, Palynological Laboratory, Solna, Sweden I express sincere appreciation of the instructions that, for the past thirty years, he has never failed to supply. My thanks, too, to Dr. Alfred R. Loeblich, Dr. Warren Drugg, and Mr. George McDougal, all of the Chervon Research Company; Dr. W. B. Critchfield, Institute of Forest Genetics, Placerville; Dr. K. K. Ching, Forest Research Center, Oregon; Professor Kenton L. Chambers, Department of Botany, Oregon State College; Professor John R. Haller, University of California, Santa Barbara; Dr. C. V. Morton, Smithsonian Institution, Washington, D.C.; Professor L. Benson, Pomona College; Professor Mildred Mathias, Department of Botany, University of California, Los Angeles; Professor L. R. Wilson, University of Oklahoma; Professor John M. Tucker, University of California, Davis; Dr. Luther W. Norem and Dr. G. Kiss, Richfield Oil Corporation; Dr. C. C. Woo, U. S. Geological Survey, Washington, D.C.; Professor Jane Gray, University of Oregon, Eugene; Professor Paul Martin, University of Arizona, Tucson; and Professor Donald Whitehead, Williams College, Williamstown, Mass., for their kindness in supplying me with reference material, peat and pollen samples, slides for testing, information on certain laboratory techniques, identification of herbarium specimens in several doubtful cases, and discussions of the subject matter. The interest they showed in this research provided much encouragement.

I am also grateful to Dr. James Mosimann, National Institute of Health; Professor S. Y. Wu, Iowa State University; Professor S. T. Hu, Department of Mathematics; Professor M. R. Mickey, Jr., and Mr. D. R. Goyette of the Biostatistics Unit, School of Medicine, and Mr. S. F. Liu, Department of Engineering, University of California, Los Angeles, for their evaluation and analysis of the statistical data and suggestions related to the computer programming of the material. Finally, the many friends and colleagues who have contributed to this work in one way or another, I thank deeply.

This investigation has been generously supported by grants from the National Science Foundation. I acknowledge with gratitude the encouragement by this institution.

Part I

POLLEN MORPHOLOGY

In a search for criteria for specific identification of pine pollen it is natural to begin with the morphological features, which may be seen under the microscope. Time and again certain features have been noted and attempts made to employ them as means of differentiation, either individually or in combination. These attempts have not been successful because the features are not constant enough to be of specific value, though some are quite uniform at the subgeneric level. Other features may be of value provided they can be expressed quantitatively, for the distinction between species rests on such minute differences that a qualitative diagnosis would be inaccurate.

Features frequently mentioned by other investigators, and employed for one reason or another, may be listed as follows:

1. Relief of the distal membrane.
2. Pattern of cappa and sacci.
3. The characters of the leptoma.
4. Exine stratification.
5. Type of sacci attachment.
6. Peculiar features such as (*a*) triradiate markings, (*b*) crests at roots of sacci, and (*c*) shape of the corpus.
7. Abnormality, including (*a*) aberrant grains, (*b*) abortive grains, and (*c*) freaks.
8. Other quantitative measures such as angles between sacci and corpus, area or density of pattern on corpus or saccus, and other dimensions of a pollen grain which can be measured.

I have investigated all these items, some in greater detail than others, and found that the relief of the intersaccale distal membrane of the corpus is a very consistent feature and can be used in subspecific groupings of the pines. I also found the pattern of the cappa a useful guide in distinguishing pollen grains of some pine species from others. Other features listed may be of value, but only in a limited way (Ting, 1965).

The Distal Membrane of Corpus

The corpus of a saccate pollen grain is subdivided into two parts. The cappa on the proximal half is covered with a sexinous pattern. The distal half is the cappula usually without relief. The cappula actually occupies the aposaccale area and parts of the mesosaccale areas of the pollen grain (fig. 1).

In pollen grains of some species of pine, the intersaccale distal membrane, where the leptoma is situated, is dotted over with elements of relief. An LO-analysis shows that these "dots" are excrescences of varied size and shape. Some are granular, some are verrucose, and some are irregular ridges. These "dots" also occur on distal membranes of other saccate pollen grains in the *Pinaceae* (*Abies, Picea, Tsuga, Cedrus, Keteleeria,* and so on), and in some nonsaccate gymnosperms as well.

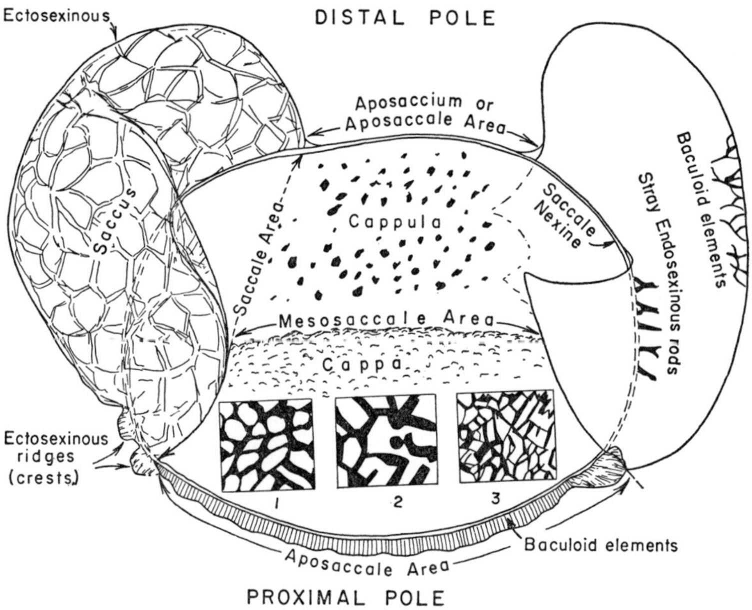

Fig. 1. A diagrammatic view of a pine pollen. The patterns illustrated on the cappa are from pollen grains of different species: (1) *Pinus flexilis;* (2) *P. jeffreyi;* (3) *P. coulteri.* Scale × 1,800. The relief shown on the cappula is typical of Group A pines.

The distal membrane of pine pollen grains is nexinous in nature. The excrescences are either superficial elements formed only on the nexinous membrane at the distal surface, or are rudimentary sexinous elements.

Since this relief feature, which here is called Type A relief, occurs consistently in pollen grains of eight species of *Pinus,* their presence becomes the basis of subgrouping these eight species into one group—Group A—and the others, without such relief, into Group B. The Group A species are: *P. albicaulis, P. flexilis, P. lambertiana* (pl. 1, fig. 1), *P. monticola, P. aristata, P. balfouriana, P. monophylla,* and *P. torreyana.* All these pines have 5 needles except *P. monophylla.* All the 2- and 3-needled pines (pl. 1, fig. 2) have pollen grains belonging to Group B.

Pattern of the Cappa

The proximal part of the corpus, or the cappa, is tectate and crassiexinous. In sectional view the ectosexine is tectate and is supported by densely spaced baculoid elements. The surface of the tectum is undulating, having depressions of various sizes, and is often perforated by micropuncta and "funnels." The supporting baculoid elements are not always simple. Ultrathin sections of the pollens of *Pinus balfouriana* show branching of these elements both upward and sideways. The upward branches support the tectum resembling a groin vault, whereas the horizontal extensions seem to fuse together forming a ceiling, to which the term "tignum" has been given (Ting and Tseng, 1965). In plan view, the baculoid elements and the tignum give a superimposed image of an infratectate reticuloid pattern. Two patterns may be distinguished as follows:

W pattern. Pattern is composed of short segments of muri, more or less uniform in size and shape. They are rather densely spaced and the pattern appears "beaded" (pl. 1, fig. 3). This pattern is found on the cappa of pollen grains of *P. albicaulis* (pl. 2, fig. 1), *P. flexilis* (pl. 2, fig. 2), *P. monticola* (pl. 3, fig. 1), *P. monophylla* (pl. 3, fig. 2), *P. muricata* (pl. 7, fig. 3), *P. radiata* (pl. 4, fig. 1), *P. remorata* (pl. 5, fig. 1), *P. torreyana* (pl. 7, fig. 4), and most grains of *P. sabiniana* (pl. 4, fig. 3).

Y pattern. Pattern is composed of longer, winding muri of irregular size and shape. The pattern is more ornate in appearance than reticuloid, and resembles the graphic texture in granite (pl. 1, fig. 4). It occurs on cappa of pollen grains of *P. jeffreyi* (pl. 5, fig. 3), *P. ponderosa* (pl. 5. fig.2), *P. coulteri*, *P. aristata* (pl. 6, fig. 2), *P. balfouriana* (pl. 6, fig. 1), *P. contorta*, *P. c.* var. *bolanderi* (pl. 6, fig. 3), *P. murrayana* (pl. 6, fig. 4), and *P. lambertiana* (pl. 3, fig. 3). The hybrids relating to the first three species, and the varieties of *P. ponderosa,* including *P. washoensis,* also have the Y pattern.

P. attenuata appears to have both patterns. Some grains of *P. sabiniana* also appear to have the Y pattern instead of W, although it is listed under W above. Distinguishing between the two patterns may be difficult in borderline cases, hence the feature is used as the basis for distinction only when it is unmistakable.

Crests and Sacci Attachment

The sexine of the cappa becomes detached from the nexine of the saccale area, and the ectosexine expands to form the sacci. The endosexinous element is also seen beneath the outer wall of the sacci, as well as over the saccale nexine. That the saccale wall is formed by nexine seems to be the prevailing interpretation.

While in dry state, the sacci are drawn together, covering the distal area, and are evidently small in size. Between the distal roots of the sacci, the sexine (ectomesine or endosexine?) is thin and is also loosely attached to the nexine, except at the roots. This thin sexinous membrane serves the function of drawing the sacci together. That it is elastic can be deduced from the variety of forms the membrane retains after the sacci open and expand. The opening of the sacci applies a tensional stress to the membrane, which either breaks or stretches to form a band between the sacci (fig. 2:2, pl. 1, fig. 2). If it breaks, its vestiges shrink and can be seen as hornlike protrusions at the bases of each of the sacci (fig. 2:3). The ectosexine is elastic because the sacci are capable of expanding from their initial much smaller size even after the grain is mounted on a slide. Whether or not the endosexine is also elastic is not clear, for if the strayed rods observed beneath the outer wall of the sacci are endosexinous, it evidently has been torn apart from the saccale wall indicating its more rigid nature.*

The elastic nature of the ectosexine seems to have cytological bearing on the origin of the crests, the often prominent undulations on the proximal magin of the corpus of some pine pollen grains, and the attachment of sacci on corpus. These three features are related. The basal circumference of the saccus is firmly attached to the corpus. The difference between *Haploxylon* and *Silvestris* types (Rudolph,

* An excellent contribution regarding exine plasticity was published while this paper was being proofread (see Banerjee, Rowley, and Alessio, 1965).

1935) lies in the point of contact of the sacci base with the corpus. If the diameter of the circular base of the sacci between proximal and distal roots is shorter than the height of the corpus, the attachment is the *Silvestris* type. If they are equal or larger, the attachment is the *Haploxylon* type. It follows that the saccus in a *Silvestris* type is attached in the distal half of the grain, whereas in the *Haploxylon* type it is more proximal.

The attachment has a direct bearing on the development of crests. When the sacci grow, their proximal bases thrust against the side of the corpus, producing folds over the proximal aposaccale area. The ectosexine may even become de-

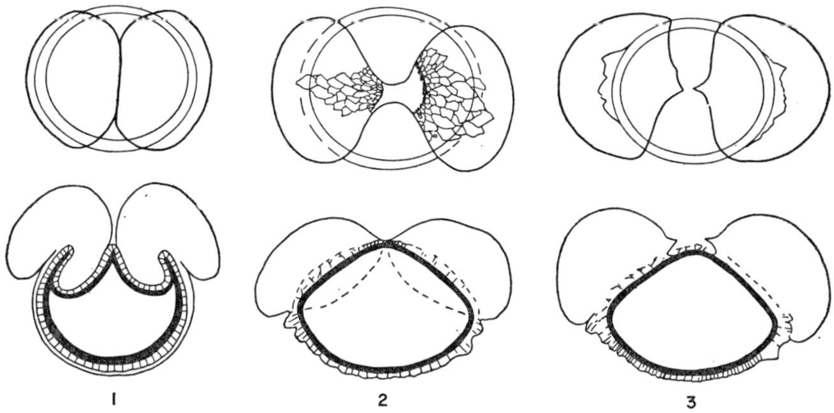

Fig. 2. Interpretation of the exine stratification. Both equatorial and polar views are shown. (1) Pollen with sacci unexpanded and folded. (2) Pollen expanded. Note connecting band formed of endosexine between sacci and occurrence of crests. (3) Pollen completely expanded. The connecting band is broken, exposing nexine (solid black) with remnant of baculoid elements on nexinous distal membrane.

tached owing to compressional stress. In that case, a very prominent crest will be formed (fig. 2). The *Haploxylon* type usually produces low crests, whereas the *Silvestris* type is apt to produce higher ones. The fact that both high- and low-crested grains occur within the same species renders the feature unsuitable as a criterion for identification.

When the knowledge concerning the distal membrane relief came to light, the pollen grains with such relief were found to coincide with the phyletic *Haploxylon* pines (Ueno, 1948, pls. 107, 174, fig. *g;* see also illustrations in Van Campo, 1950, figs. 5, 33, 34, 57; and Erdtman, 1957, fig. 55). My attempts to correlate attachment of sacci with taxa have not shown any consistent results. So far as I know, there is complete agreement neither between type of sacci attachment and distal membrane relief Type A, nor between type of sacci attachment and taxa.

Abnormal Pollen Grains

Normal pollen grains of *Pinus* are disaccate; the sacci are approximately equal in size. The corpus and the sacci patterns are always well developed. Abnormal pollen grains are of three kinds (fig. 3; pl. 7, figs. 1, 2):

1. Aberrant grains are either multisaccate or monosaccate, and very rarely have a bicellular corpus.

2. Abortive grains always have dwindled sacci that are devoid of pattern.

3. Freaks are those where the size of the entire grain or any component part of it is either unusually large or small.

The aberrant grains appear to be of regular occurrence among all pine species. They do not exceed 3 per cent. The abortives do not occur in all species nor in all samples of one species, but when present are more abundant. The freaks are rare and never exceed 0.3 per cent (Srivatava, 1961, gives 0.18 per cent for *P. roxburghii* in India). No correlation has yet been noted between taxa and abnormal pollen grains. Studies of some samples of hybrids do not show a high frequency of aberrant grains, but may have numerous abortive grains.

Shape and Orientation of the Pollen Grain

In figure 4 various shapes of pine pollen grains are shown. The general shape of the corpus is ellipsoidal. Its longer axis is parallel to the overall breadth of the grain. But the grains of *P. lambertiana* frequently deviate from this plan. Many of them are angular, tending to hexagonal in shape, and have their longer axis at right angles to the overall breadth of the grain.

The sacci are usually hemispherical in polar view. Their base line may appear straight or concave or even convex, but their shapes are not symmetrical with respect to breadth. Pollen grains of *P. coulteri,* however, frequently have globular sacci that are symmetrical to their breadth. Some globular sacci are also found on pollen grains of other species of pine.

The overall view of a pollen grain may be highly variable. Some grains present themselves in polar views, others in lateral views. This difference in orientation may result from differences in mounting the slide, but there seems to be a high tendency for *Haploxylon* grains to be oriented in polar view, and *Silvestris* in lateral view.

Other Features

Attempts to transfer qualitative characters into a quantitative basis for identification have not met with success in the following features:

1. Attempt to define the density of pattern on cappa and on saccus has not yielded consistent results. First, it is difficult to measure the area of lumina or bronchi. Second, the number of lumina or bronchi per square unit area varies according to location. Usually the lumina become smaller outward from the proximal pole of the cappa. But this is not always true, for in grains with triradiate markings, the lumina bordering them are much finer at the proximal pole. It may be said that triradiate markings are extremely rare in pollen of *Pinus.* Nevertheless they have been seen on aberrant grains of *P. jeffreyi.* While density of pattern on cappa does enable classification of pollen grains of pine into W and Y types, the bronchi on saccus do not vary sufficiently to allow further subgrouping of the pollen grains of different pine species.

2. Measurement of the thickness of exine and that of the individual stratum of the exine may offer a good clue to distinguish some species from others. Measuring

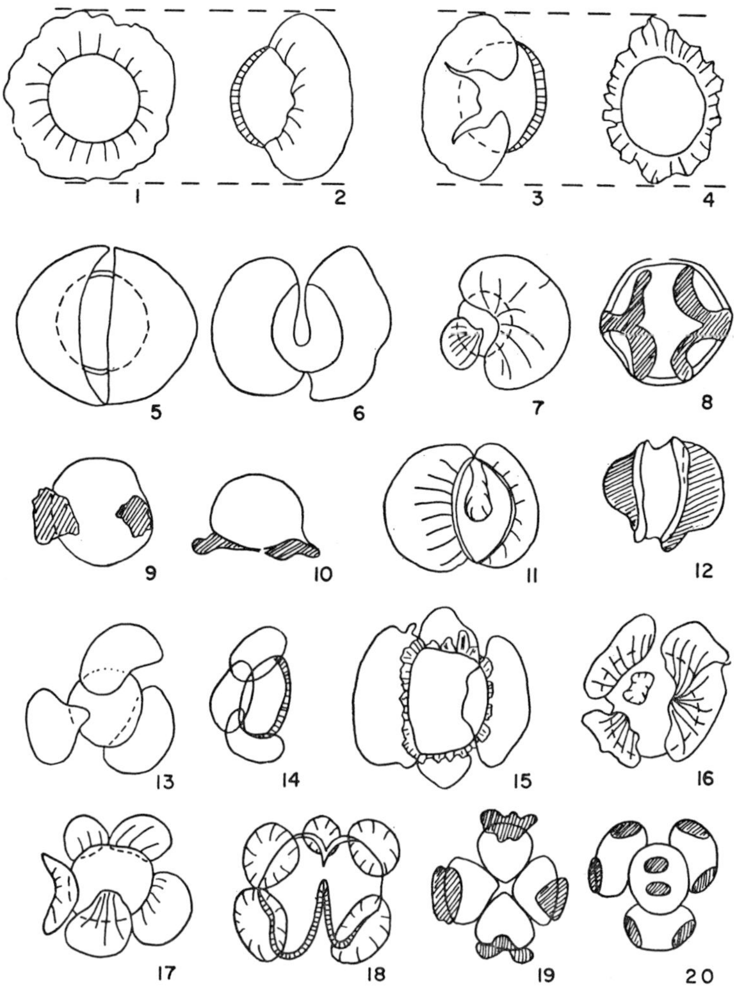

Fig. 3. Observed abnormalities of *Pinus* pollen. Sketches are not drawn to uniform scale. (1–4) Unisaccate; (5, 6) cosaccate, with one small part of the sacci united; (7–12) disaccate, very unequal in size in (7), very small in (9) and (10), tending to become trisaccate in (11); (13, 14) trisaccate; (15–17) polysaccate; (18) Siamese twins with one saccus in common; (19, 20) tetrads showing locations of sacci.

Species: (1, 2, 5, 6) *P. albicaulis;* (3, 4, 7, 14, 16) *P. attenuata;* (8, 10, 11, 12) *P. lambertiana;* (9) *P. ponderosa;* (13) *P. monticola;* (15, 17) *P. jeffreyi;* (18) *P. ponderosa* × *jeffreyi;* (19, 20) *P. muricata.*

the individual stratum is always tedious and time consuming, as it must be done under very high magnification.

3. If the area covered by the pattern is defined as coextensive with cappa, it may be found that the proportion of this area to the total area of the corpus offers a good criterion for discrimination. But I have not been able to find a good way of estimating the area. My results show that the proportion of the pattern-covered area is larger in pollen of *P. murrayana* than in pollen of *P. contorta,* and in

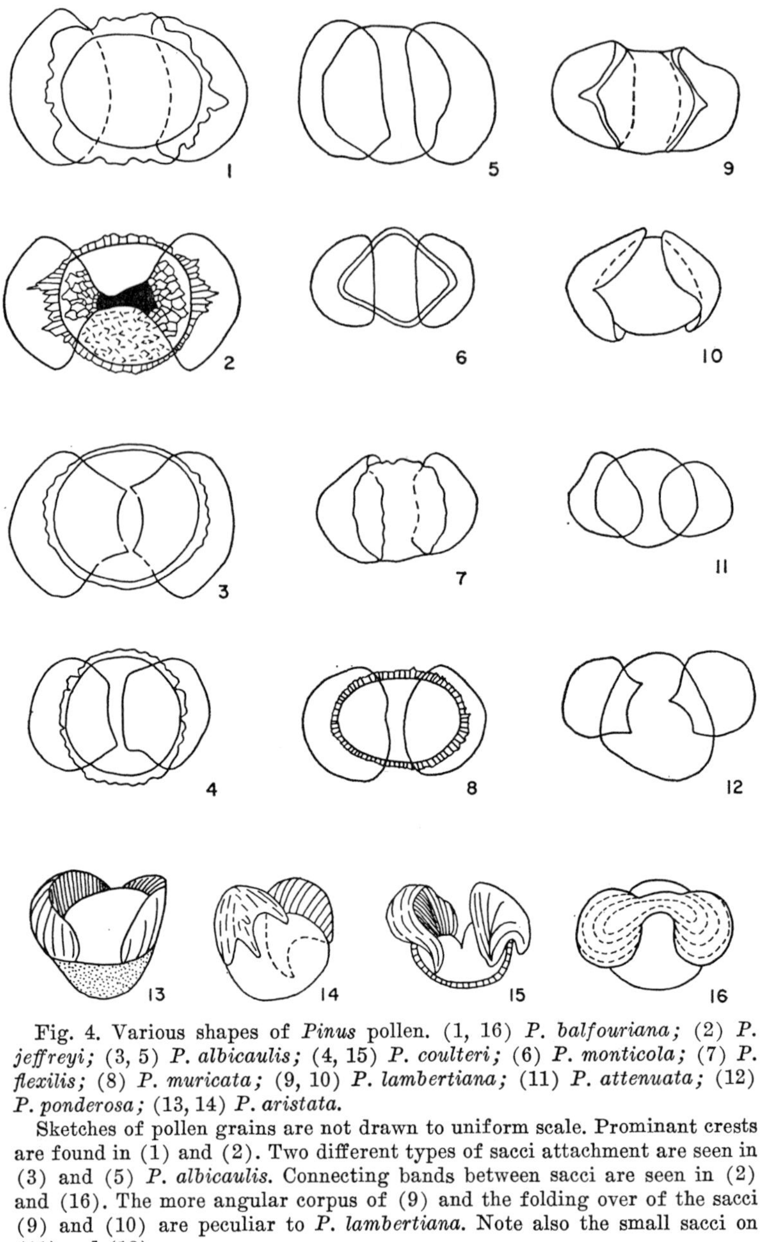

Fig. 4. Various shapes of *Pinus* pollen. (1, 16) *P. balfouriana;* (2) *P. jeffreyi;* (3, 5) *P. albicaulis;* (4, 15) *P. coulteri;* (6) *P. monticola;* (7) *P. flexilis;* (8) *P. muricata;* (9, 10) *P. lambertiana;* (11) *P. attenuata;* (12) *P. ponderosa;* (13, 14) *P. aristata.*

Sketches of pollen grains are not drawn to uniform scale. Prominant crests are found in (1) and (2). Two different types of sacci attachment are seen in (3) and (5) *P. albicaulis.* Connecting bands between sacci are seen in (2) and (16). The more angular corpus of (9) and the folding over of the sacci (9) and (10) are peculiar to *P. lambertiana.* Note also the small sacci on (11) and (12).

pollen of *P. jeffreyi* is larger than in pollen of *P. ponderosa.* Much additional work has to be done before any concrete results useful for specific discrimination appear.

TECHNIQUES

This work began in the summer of 1958. Initially the standard techniques then available were used. Fresh pollen grains were acetolyzed (Erdtman, 1943, pp.

27–28, 34–35), without chlorination and staining. Glycerine jelly was chosen as the medium for mounting. It was prepared according to Kisser, as described by Erdtman (Erdtman, 1943, p. 26). For fossil samples in various kinds of sediments, many ways have been tried. The basic technique is still that of hydrofluoric acid (HF) treatment, followed by separation with heavy liquid, then acetolyzed.

There has been a great deal of development in this field during these few years, but little of it has to do with the pollen grains of pines. In order to evaluate new techniques, new chemicals for treatment of pollen grains, and new mounting mediums, I keep testing as a constant practice.

Reference Pollen Material

Both herbarium collections and fresh pollen were used in this work. To avoid potential contamination inherent in using herbarium material, I collected my pollen samples from the microsporangiate strobilus. To ensure better approximation of a fossil assemblage, I collected strobili from different branches if available. The fresh pollen samples supplied to me were sometimes mixtures from a number of trees, while others were from single trees.

It may be desirable to obtain pollen collections from the same trees for a number of years to study their possible size variations with time. Such variation, however, may already have been covered by multiple sampling of a species, as tests based on these samples have been satisfactory.

Physical and Chemical Change in Size of Pollen Grains

My method of identifying fossil grains is not based on the measurements alone, but depends on their mutual ratios. Nevertheless, measurements differ according to the nature of response of pollen grains to varying treatments. Thus I carried out experiments on the pollen grains of pines because no detailed analysis of size changes of pine pollen grains under chemical treatment has been reported.

CHANGES OWING TO MOISTURE, HEAT, AND ACIDITY

Fresh pollen grains were desiccated (*a*) in an oven at low temperatures (150–300° F) for 10–15 minutes or (*b*) by washing with xylene and letting dry on slide at room temperature. A drop of melted glycerine jelly was added to the pollen grains on the slide and covered. This way of mounting a slide is here called a "dry" mount. The size measured (i.e., corpus breadth, $\overline{X}_1$) is considered the standard size of nontreated pollen grains.

A similarly mounted slide was then heated over an alcoholic flame until the jelly bubbled. The pollen grains immediately expand to a significantly different size from the standard (see Appendix V). This expanded size is much like that of acetolyzed samples. In other words, the expansion was directly caused by heat. The heat probably caused moisture in the cell to vaporize and thereby to exert pressure on the cell wall. The wall must be elastic so that it stretches and yet retains shape. There must also be some plastic deformation so that the expanded shape is retained in the mounting medium.

In the acetolyzed samples, both heat and acidity are involved. Christensen (1946, p. 15) describes the expansion as taking place "so suddenly and completely . . . that it was terminated before the sample even came into the water

TABLE 1

EXPANSION OF POLLEN SIZE OF *Pinus ponderosa* 17

(In μ)

Time	Grain 1			Grain 2		
	$\overline{X}_1$	$\overline{X}_2$	$\overline{X}_3$	$\overline{X}_1$	$\overline{X}_2$	$\overline{X}_3$
24 hours	49	43	72.0	60	47.5	86
29 hours	50	43	72.5	60	47.5	86
48 hours	51	43.5	72.5	60	47.5	86
1 week	51	43.5	72.5	60	47.5	86
8 days	51	44	73.0	61	48.0	86
21 days	51	44	73.0	61	48.0	86
368 days	52	45	74.0	60	50.0	89

TABLE 2

EXPANSION OF POLLEN SIZE IN GLYCERINE-JELLY MOUNTS

(In μ; $N = 100$ grains)

Species and sample number	Slide	Time	$\overline{X}_1$	SE	$\overline{X}_3$
P. attenuata 29	1 (O)[a]	1 day	64.34	2.88	
		2 years	66.05(−)[b]	1.77	
P. attenuata 35	1 (O)	1 day	68.48	1.94	
		2 years	70.50(−)	2.09	
P. attenuata 35	2	2 years	72.58(+)[c]	2.01	
	3	2 years	74.12(+)	1.88	
	4	2 years	73.46(+)	2.32	
P. ponderosa 1	1 (O)	1 day	61.85		87.33(+)
		240 days	70.40(+)		95.52(+)
P. jeffreyi 26	3*a*(O)	1 day	77.22	2.37	
	3*b*	60 days	80.06(−)	2.07	
	3*c*	90 days	81.56(+)	2.31	

[a] (O) Original basis for comparison.
[b] (−) Not significant.
[c] (+) Difference statistically significant from original basis (O).

bath." The suddenness of the response suggests an extremely rapid chemical reaction which generates heat.

Pollen embedded in glycerine jelly expands in size with time (tables 1, 2). I surmised at first that moisture content of the mounting medium was responsible for the gradual increase in size, but I am now inclined to believe that it is because of the acid reaction to the acidic content of the glycerine jelly.

Desiccated pollen grains mounted by adding a drop of water before adding the melted glycerine jelly (i.e., "wet" mounts) also expanded in size. The expanded size is not significantly different from that of an acetolyzed sample.

We now have three different slides: (*a*) fresh desiccated pollen grains in dry mounts, but heated; (*b*) fresh, desiccated pollen grain in wet mounts, only gently heated to melt the jelly; and (*c*) acetolyzed pollen grains in wet or dry mounts. All give essentially the same mean size. The cause for expansion is vaporization of latent moisture owing to intense heating, either directly applied or generated by chemical reaction.

ELASTICITY OF POLLEN WALL

Pollen size is a function of the elasticity of the pollen wall and chemical environment. There is a predisposed response to every chemical, and possibly a certain size pertinent to that change. Little or no expansion will occur by overheating an acetolyzed sample mounted in glycerine jelly, since the response to acetolyzed heat has already been achieved.

Regarding the pollen wall, the data in Appendix V show a significant expansion of size in both breadth of saccus and overall dimension. The sacci are enlarged more than the corpus because the elasticity of the sexine is higher than that of the nexine. The fact that the saccus is formed by a part of the sexine which is not strongly rooted to the nexine, as is the sexine of the corpus, also contributes to the greater expansion of the saccus. The significant expansion of the overall breadth owes to the expansion of the sacci which are now openly extended instead of closely retracted over the distal pole.

That the pollen wall retains its elastic property for a long time is evident because (*a*) fossil pollen grains change their size in glycerine-jelly mounts, (*b*) fossils from Early Pleistocene are found to react differently to HF and acetolysis (AC), and (*c*) fossil as well as modern pollen grains change their size when transferred from one chemical medium to another (Andersen, 1960; see also Appendix VI).

COMPARISON OF SIZES OF FOSSIL AND MODERN POLLEN GRAINS

To identify fossil pollen grains by comparing their size with known modern grains, the assumption must be made that by using the same chemical treatment, a fossil pollen grain will respond in much the same way as a modern grain. For a statistical analysis, the assumption applies not to an individual grain, but to the mean size of a number of grains. Since there are different methods of treating pollen, these were investigated. Data of Christensen (1946) and Andersen (1960) regarding *Corylus avellena* have become the standard measure for judging the normality of other pollen grain sizes in a preparation (Faegri, 1964, p. 28). Their results are discussed in Appendix VI.

Table 3 compares the data by Christensen (1946) and Andersen (1960) with my results. The table shows the sizes of pollen grains after treatment and their relative relationships in terms of significant difference. Christensen's data show that fossil pollen grains are smallest when treated with HF and larger when treated with potassium hydroxide (KOH), but still smaller than the nontreated

TABLE 3

SUMMARY OF COMPARISONS OF POLLEN SIZE AFTER VARIOUS TREATMENTS

Sample	Author	Pollen tested	Comparison of size after treatment (arranged in ascending order, from small to larger, left to right)
Fossil	Christensen, 1946	*Corylus avellena*	HF <[a] KOH < Nontreated < HF + AC < AC ≤[b] KOH + AC
	Ting (data compiled 1962–1964)	*Pinus*, *Tsuga*, *Betula* from peat of Maine	$\{$Nontreated; =[c] KOH; = HF$\}$ < $\{$AC; = HF + AC$\}$ < KOH +[d] AC
	Andersen, 1960	*Corylus avellena*	KOH < KOH + HF + AC < KOH + AC
Modern	Christensen, 1946	*Corylus avellena*	$\{$Nontreated; = KOH$\}$ < AC ≤ KOH +AC
	Ting (data compiled 1962–1964)	*Pinus jeffreyi* X *ponderosa*	HF < $\{$Nontreated; = KOH$\}$ < HF + AC + HF < $\{$AC; = HF + AC$\}$
		Size: $\overline{X}_1$ and SD, in μ	58.04 ± 3.32 < $\{$65.90 ± 4.48; 6737 ± 4.48$\}$ < 72.36 ± 5.25 < $\{$75.38 ± 6.37; 76.25 ± 7.64; 76.49 ± 5.84; 75.44 ± 5.07; 77.18 ± 5.29$\}$
Modern versus fossil	Christensen, 1946	*Corylus avellena*	HF (*f*) < $\{$KOH (*f*); = nontreated (*f*)[e]$\}$ < $\{$KOH(*m*); = nontreated (*m*)[f]$\}$ < $\{$AC (*m*); = HF + AC (*f*)$\}$ < KOH + AC (*m*) < AC (*f*) < KOH + AC (*f*)
		Size: $\overline{X}_1$ and SD, in μ	21.0 < $\{$23.4; = 22.7$\}$ < $\{$25.7; = 25.7$\}$ < $\{$28.0; = 27.8$\}$ < 31.4 < 33.9 ≤ 34.4

[a] < Significancantly different in mean. For example, HF < AC indicates that the size treated by HF is smaller than the size treated by acetolysis.
[b] ≤ Equal or significantly different in mean.
[c] = No significant difference in mean.
[d] + If samples receive more than one treatment, the different treatments are connected by +. The treatment before the sign comes first, the treatment after the sign later (i.e., HF + AC + KOH indicates the sample has undergone three successive treatments).
[e] (*f*) Fossil.
[f] (*m*) Modern.

standard. Those treated with (*a*) HF + AC, (*b*) AC alone, and (*c*) KOH + AC are successively larger, and all are larger than the standard. By contrast, my investigation shows that HF- and KOH-treated fossil pollen grains of *Pinus, Tsuga, Betula,* and *Fagus* (partially) do not differ in size from the standard. The acetolyzed grains are similar in size to those grains treated by HF and AC. In modern pollen grains, Christensen's data show that acetolyzed *Corylus* grains are smaller than KOH and acetolyzed grains. My data show that acetolyzed *Pinus* grains are smaller than KOH and acetolyzed grains, but are not significantly different from HF and acetolyzed grains.

Since KOH treatment was not used in my work (see Appendix VII), and since both fossil and modern pollen grain sizes of acetolyzed samples are not significantly different from those treated with HF and acetolyzed, it is possible to compare the fossils with modern references so treated. HF treatment makes no difference in size of fossil pollen grains if these are subsequently acetolyzed. In fact, when one compares the significance of mean values of both fossil and modern grains of Christensen's data, one sees that the modern acetolyzed pollen of *Corylus avellena* is not significantly different from fossil *C. avellena* treated by HF and acetolysis.

TABLE 4

SIZE OF POLLEN GRAINS AFTER ACETOLYSIS

Lot	*N*	AC treatment (mixture of acetic anhydride and sulphuric acid)	$\overline{X}$ (in μ)	2SE
A.	100	9:1 mixture	57.62	1.12
B.	100	9:1 mixture	56.60	1.19
C.	100	9:2 mixture	56.60	1.07
D.	100	In 9:1 mixture for 2 weeks after boiling	59.11	0.99
Average sample of *A* + *B* + *C*.	300		57.04	0.92

SOME INVESTIGATIONS REGARDING ACETOLYSIS

Since high acidity appears to increase the elasticity of grains, I investigated this feature. The standard acetolyzed solution contained one part of sulphuric acid, H_2SO_4, to nine parts of acetic anhydride, $(CH_3CO_2)O$. I increased the amount of sulphuric acid to two parts and found no appreciable difference in pollen size (table 4, lot *C*).

If the grains are left in an acid solution for a long time their size will be affected. Christensen (1946) noted the *decreased* size of pollen grains of *Corylus avellena* in acetolyzed solution while the samples were left standing in the water bath (see Appendix V). My experience with *Pinus* pollen grains is that they *increase* in size very significantly when left in acid for two weeks. The grains may have diminished in size to begin with, though I consider this as probably simple variation between two random samples. Of course the length of time the pollen grains are left in acid may make all the difference, for I left the pollen grains in acid much longer than Christensen did (table 4, lot *D*).

Acetolyzed pollen grains treated, if not thoroughly washed, will expand in size in storage. Samples stored in water, glycerine, or 70 per cent alcohol will also show some expansion after some time. To wash off completely the acetolyzed solution from pollen samples, one needs 15 to 20 times as much distilled water as the original quantity of acetolyzed solution, after centrifuging and decanting the supernatant liquid. Three or four rinses in a 15-cc tube is clearly not sufficient. For good storage, absolute alcohol should be used, and the tube should be refilled each time it is opened.

TABLE 5

FOSSIL POLLEN OF *Pinus* AFTER VARIOUS TREATMENTS

Lot	Treatment	$\overline{X}_1$ (in μ)	2SE	$\overline{X}_1$ of lots compared	Proportion of change
A	AC only	61.97	1.36	*A to B*	101
B	Sample *D*, acetolyzed	62.48	1.29	*B* to *D*	117
C	In HF for 10 months	47.36	1.02	*A* + *B* to *D*	117
D	In HF for 4 hours	53.36	0.90	*B* to *C*	132
Average of sample *A* + *B*		62.24	0.93	*C* to *D*	112

NOTE: Fossils from Merced Formation (Early Pleistocene) in California.

THE USE OF HYDROFLUORIC ACID AND HEAVY LIQUIDS

Hydrofluoric acid (52 per cent) and heavy liquids are used to remove mineral matter in sedimentary samples. The former is a chemical process, the latter a physical process. Hydrofluoric acid is used to remove silicates. There are three ways of using this reagent:

1. Maceration by prolonged soaking of sediments in HF.
2. Short treatment by heating sample in acid in a water bath or in a sand bath.
3. Short treatment by pouring boiling HF over sediments.

Treatment with HF diminishes the size of the pollen grains. The longer the maceration, the smaller will be the pollen grains. Comparison of a sample treated by boiling four hours in HF with a sample soaked in acid for ten months shows that the latter treatment gives a significantly smaller size than the former (table 5). This applies equally to recent and fossil pollen grains of *Pinus*.

When a previously HF-treated sample is subjected to acetolysis, the pollen grains react as if they had not been HF treated. Their sizes increase, but not to any extent significantly different from those treated by acetolysis only (Appendix VIII).

Sample treated by HF should be thoroughly washed, at first with water to which have been added a few drops of HCl, then by many rinses with water.

Better still is to eliminate carbonates prior to HF treatment by HCl treatment. This will eliminate the problem of having precipitates of calcium fluoride after HF treatment.

Silica may also be removed by concentrated KOH under an infrared source of heat. I did not use this procedure because I have avoided the use of KOH.

Mineral matter other than silicates may be chemically dissolved by other reagents, but I have not employed them. The less one subjects the pollen grains to chemical process the better will be the chance of maintaining uniformity in pollen size and preservation.

Density separation by means of panning under low magnification is used by McDougal of the Chervon Research Company. I have always used heavy liquid flotation to remove the mineral particles in samples with or without HF treatments.

The medium first used was bromoform (tribromomethane, $CHBr_3$, specific gravity 2.87 at 20° C) diluted with alcohol (C_2H_2OH, specific gravity 0.8) to a specific gravity of 2.0. Later I tried zinc chloride solution ($ZnCl_2$, specific gravity 1.96). The high viscosity of zinc chloride solution is its major defect, though it is less toxic and less expensive than bromoform. I suspected also that zinc chloride may have a diminishing effect on pollen grain size, but I did not make any tests to find out about this point.

Acting on the suggestion of Professor James D. McCullough of the Chemistry Department, University of California, Los Angeles, I tested another agent for separation. It is dichlorodibromomethane (CBr_2Cl_2), an agent with suitable specific gravity (2.42) and polarity. It does not cause clumpiness. It is soluble in alcohol. Dehydration of samples is required, although a slight quantity of moisture is miscible. Samples that did not yield good results in bromoform give good concentration of pollen in this liquid. Its only drawback is its costliness.

The separation by means of density must be done in a number of steps. Tests show that saccate pollen grains are separated out better in a liquid with a higher specific gravity, such as 1.8 or 2.0, whereas pollen grains of angiosperms are better concentrated with a liquid specific gravity of 1.4–1.5 (table 6). The specific gravity of acetolyzed pollen grains is also different. Pollen grains of *Abies, Pinus,* and *Picea* have a higher density than those of angiosperms, some of which may have as low a density as 1.45. It is therefore necessary to subject the sample to a series of separations by flotation in liquids of increasing densities, through steps such as 1.4, 1.6, 1.8 to 2.0 (Appendix IX).

Mounting Media and Size Changes of Pollen Grains in Glycerine-Jelly Mounts

I have used glycerine jelly for mounting slides from the outset, and in connection with this project still do. I investigated a few other mounting media. One of these was Diaphane, which I found to be unsatisfactory. Acetolyzed pollen grains of *P. flexilis* mounted in Diaphane produce a smaller size than those mounted in glycerine jelly, by these amounts: corpus, 32.5 per cent; saccus, 37.0 per cent; and overall breadth, 15.1 per cent. There is also a tendency for the pollen grains to diminish in size as the Diaphane hardens. Fading of important morphologic

TABLE 6

DENSITY OF POLLEN GRAINS OF VARIOUS TAXA MEASURED BY SEPARATION, USING BROMOFORM

Species and sample number	Density when pollen grains are floating throughout the liquid from top to bottom after centrifuging	Density when pollen grains settle to bottom of tube after centrifuging
Abies concolor 1A	1.50	1.45
Abies concolor 1B	1.55	1.55
Abies concolor 1C	1.55	1.50
Abies magnifica 2	1.50	1.45
Picea sitchensis 1A	1.50	1.45
Picea sitchensis 1B	1.60	1.55
Pinus albicaulis 13	1.45	1.40
Pinus aristata 17	1.45	1.40
Pinus aristata 10	1.45	1.40
Pinus attenuata 30	1.50	1.45
Pinus balfouriana 12	1.45	1.40
Pinus balfouriana 18	1.45	1.40
Pinus balfouriana 4	1.50	1.45
Pinus contorta var. *bolanderi* 7	1.50	1.45
Pinus contorta 16	1.65	1.55
Pinus coulteri 8	1.45	1.45
Pinus flexilis 5	1.50	1.45
Pinus jeffreyi X *coulteri* 21	1.70	1.65
Pinus lambertiana 9	1.40	1.40
Pinus monophylla 7	1.60	1.55
Pinus monticola 13	1.60	1.55
Pinus muricata 13D	1.45	1.45
Pinus murrayana 2	1.50	1.45
Pinus ponderosa 32NA	1.50	1.45
Pinus ponderosa 32NB	1.50	1.50
Pinus ponderosa 32NC	1.55	1.55
Pinus radiata 7	1.70	1.65
Pinus sabiniana 11	1.50	1.45
Pinus torreyana 4A	1.50	1.45
Pinus torreyana 4B	1.45	1.40
Pseudotsuga macrocarpa	1.45	1.40
Pterocarya stenoptera	1.60	1.50
Rhododendron indicum	1.50	1.40
Sidalcea malvaeflora	1.50	1.40
Dicentra chrysantha	1.50	1.50
Vicia mexicana	1.50	1.45
Madia elegans	1.50	1.40
Eriophyllum sp.	1.50	1.40
Gilia sp.	1.60	1.50

features seems to be the greatest drawback to Diaphane. Another medium I tried was Dupont Cellosize after fixation by Evanol. Acetolyzed grains of *Picea sitchensis* from Oregon showed a difference of corpus size by 5.9 per cent, the glycerine-jelly–mounted pollen grains being the larger.

The reference measurements I made for pollen grains were all based on slides mounted with glycerine jelly. The pollen count for fossil samples rich in pollen grains, and for fresh pollen samples can be made simply from a water mount. Preservation and observation of certain pollen grains are possible from single-grain slides. Pollen grains measured in water do not differ significantly in size from those in glycerine mount (see Appendix VI).

Pollen grains mounted in glycerine jelly, whether recent or fossil, tend to increase in size with time (tables 1, 2). The rate of increase is erratic, but in no case did the change become significant within the first three days of mounting, and practically no change of size occurred within the first 24 hours. I always took my measurements as soon as the jelly sets. My rule is to make a new slide whenever measurements are needed. I do not use a slide warmer lest the constant low, but prolonged, heating cause expansion of pollen grains.

Slides are mounted thick so that pollen grains will not be compressed by the weight of the cover glass (Cushing, 1960). Place a drop of melted glycerine jelly on the pollen material at the middle of a cover glass, and stir and mix until even. It will set in a minute or two. Meanwhile, place a few granules of paraffin on the four corners of the cover glass and press slightly until the paraffin sticks to the glass. Turn the cover glass over and place it on the slide in the position desired. Heat it gently, and the paraffin and jelly will melt and spread. The mounting and sealing of the slide are completed simultaneously.

Part II

STATISTICS

Measurements

The use of measurements as criteria for specific differentiation of pollen has a long history. The in-species variation, however, as well as the considerable overlap in size among grains of different species, has been of constant concern. To many, the obstacle has been insurmountable. Others have developed complete confidence in frequency curves and employ them as the primary means for identification.

Measurements of certain dimensions of a pollen grain constitute the basic procedure of this study (fig. 5). Pine pollen grains exhibit various forms and shapes on a slide. They may be oriented in the polar view, in the equatorial view, or in any oblique view. The saccus may be fully extended in open view or may be infolded in a closed view. Previous workers have used different dimensions as a basis for measurement, and most have used only one of the several possible. I used three (see fig. 5): breadth of corpus (C), breadth of the saccus (S), and overall breadth of grain (O) (Erdtman, 1957, p. 6). Dimension S is perpendicular to O and C and is therefore independent of O and C. Dimension C is, in most cases, parallel to O and is therefore not independent of O.

The pollen grains are measured on a mechanical stage column by column. Each grain that will give the one or more desired dimensions *must* be measured until an established sample size has been reached. The individual values of measurements of C, S, and O are then totaled and the mean values of the measurements of each dimension are found. Ratios are then calculated from these mean values.

Only the data of disaccates with well-exhibited patterns on both corpus and saccus are used as reference material. Abnormal pollen grains are excluded. Though not used as references, their frequence of occurrence and their dimensions are also recorded.

The breadth of corpus as defined by Erdtman (1957, p. 6) does not include the thickness of the exine. My measurements, however, have included this. Only in extreme cases, where prominent crests are developed, is there some difficulty in ascertaining the normal thickness of exine. I have, in these cases, added twice the thickness of exine, measured from the proximal pole, to a measurement of breadth of corpus from its inner margins. The reason for measuring from the outer rim of the corpus is to take into consideration the possible diagnostic value of the thickness of exine.

The Scattergram and Sample Size

I used the scattergram for identification purposes in my work of 1958. It was termed the distributional diagram of dimensions (Axelrod and Ting, 1960, pp. 49–50). Though it has played an important role, and is still useful in cases with two dimensions, its use in the study of pollen grains of pines has now been superseded by the present method.

The diagram was based on bivariates. My first choice of the variables for pine studies involved a dependent relationship of one to the other: namely, the corpus

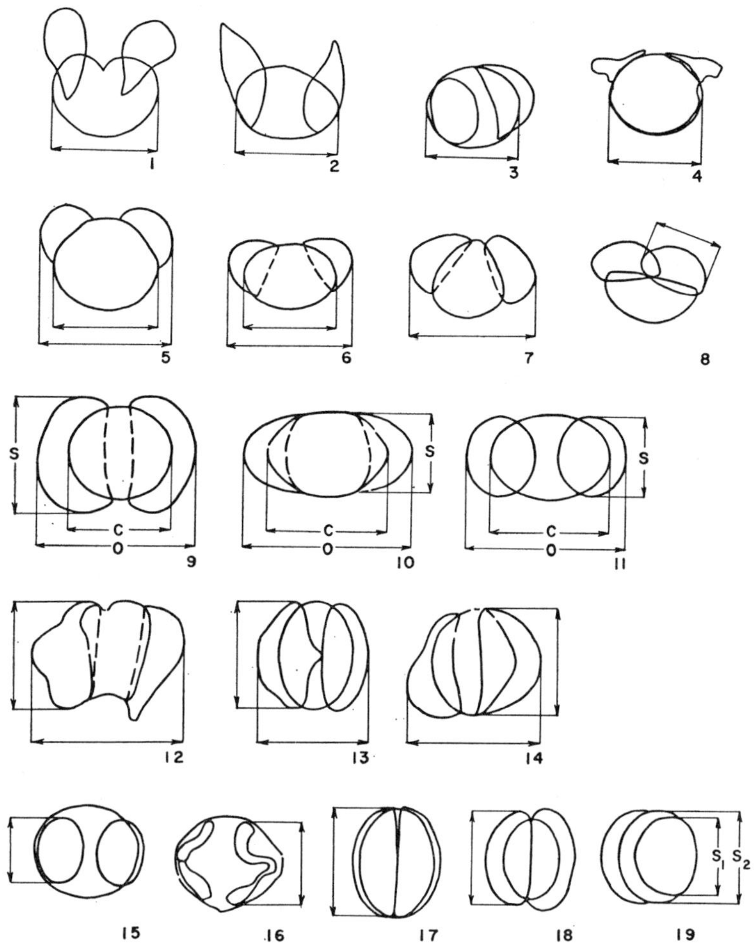

Fig. 5. Diagrams showing various orientations of pollen and dimensions measured. (1–8) Lateral view; (9–18) polar view; (9–14) open view; (15–18) closed view; (19) equatorial view. Dimensions that may be measured: (1–4) corpus only; (5, 6) corpus and overall breadth; (7) overall only; (8) saccus only; (9–11) overall, corpus, and saccus; (12–14) saccus and overall; (15, 16) corpus and saccus; (17–19) saccus only. Average of S_1 and S_2 is used in (19). The views shown in (4–6), (9–11), and (15) are preferable. Abbreviations: *C,* corpus; *O,* overall; *S,* saccus.

breadth and the overall breadth. It would be better if the bivariates were independent of each other.

The scattergram is plotted by using the two dimensions—corpus breadth and overall breadth—on rectangular coordinates (fig. 6). The diagram differs from the conventional scatter diagram in that the overall dimension is always longer than the other, and the plots are therefore limited to the lower half of the quadrant. A transformation can be achieved by using the point of mean value of both measures as the origin. Lines of equal proportion are defined by the angular distance from the $Y =$ axis since all these lines pass through the origin. The area enclosed by two isoproportional lines increases progressively with increasing dimensions. Isodimensional lines are circular arcs drawn from the origin as center with the dimension (in μ) as radius. The increment of areas bounded by successive arcs and isoproportional lines is similar to the conical projection of a globe. This means that with increase in size, the variation in the combination of corpus breadth and overall length also increases. Hence, measurements of smaller grains are more

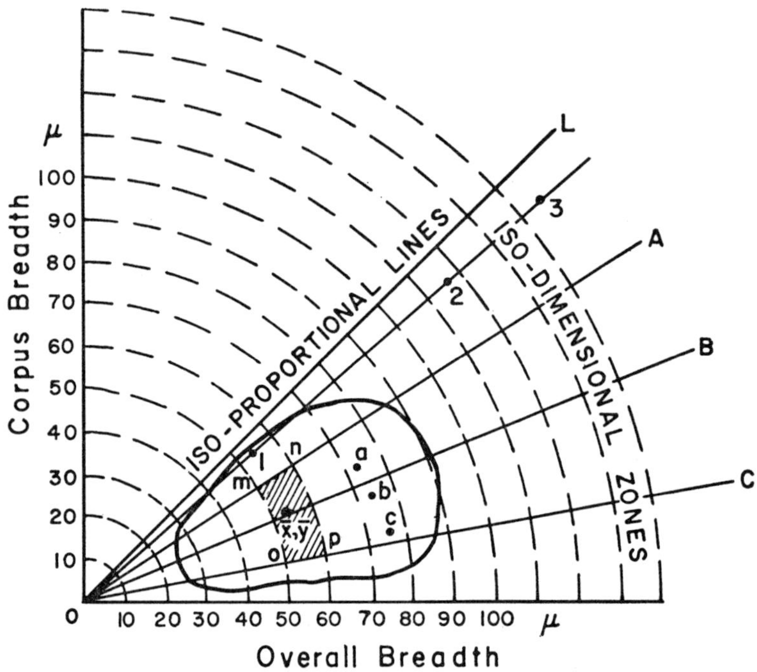

Fig. 6. Scattergram of a taxon. Encircled area is scattergram of a taxon. Area *mnop* is the central area of dispersion in which the means are located. $\overline{X}$, $\overline{Y}$ are the respective means of dimensions X and Y. Numbers 1, 2, and 3 are measurements of three pollens having almost equal proportions, but not of the same species, whereas *a*, *b*, and *c* vary in proportions but are within one isodimensional zone and are of the same species.

homogeneous than those of larger grains, and thus they give a higher correlation of coefficient between the two measurements.

The area bounded by the arcs and the lines in which the mean is located is the central area of dispersion and is used for comparison of fossil data similarly plotted. Within an area, measurements with small difference of mean and of proportion may be considered identical (see fig. 6, *a–c*).

The intervals used for the isoproportional lines in the study of angiosperms are 1–1.25, 1.25–1.5, 1.5–2.5, 2.5–3.0, 3.0 and more. For pines between 1.3–1.4 it is necessary to subdivide the second interval group (1.25–1.5) into three: 1.2–1.3, 1.3–1.4, 1.4–1.5. My use of the scattergram in early studies had the obvious advantage over the frequency curves of bringing into consideration a second dimension and its mutual correlation. By superimposing the diagram of a known species on that of an unknown, one may note the degree of coincidence between them, and judge specific similarity. The chief drawback is that the diagrams of different species seldom exclude one another. In California pines they always overlap, sometimes as much as 44 per cent. The degree of overlap is another expression of overlapping in the confidence intervals between samples.

Projections to both axes of all the points on the scattergram will give the frequency spectra of both dimensions measured.

In measuring pollen grains I kept separate lots of various sizes. Pollen grains in lot sizes of 10, 30, 40, 50, 60, and certain selected pollen grains of any lot size are

TABLE 7

SUCCESSIVE MEANS OF CORPUS BREADTH OF *P. monticola* AND OF CORPUS BREADTH AND SACCUS BREADTH OF *P. contorta* POLLEN GRAINS

(IN μ)

P. monticola		*P contorta*			
N	$\overline{X}_1$	*N*	$\overline{X}_1$	$\overline{X}_2$	Ratio $\overline{X}_1/\overline{X}_2$
10	62.51	10	56.62	41.40	1.37
20	59.16	20	55.95	41.85	1.33
30	58.88	30	55.93	41.66	1.34
40	57.67	40	56.37	41.58	1.35
50	57.49	50	55.98	41.46	1.35
60	57.29	60	55.86	41.40	1.35
70	56.57	70	56.19	41.56	1.35
80	56.75	80	56.26	41.50	1.35
90	56.30	90	56.33	41.58	1.35
100	56.27	100	56.34	41.62	1.35
103	56.54				

NOTE: Successive means of breadth of corpus of pollen of *P. monticola* showing that beyond $N = 70$ additional measurements do not materially alter the value of the mean. For *P. contorta*, $N = 40$ is sufficient.

also recorded. Only a few of these lots are listed in appendix tables 1 and 2. They are useful in determining the minimum sample size that is required for a particular sample.

While working on the scattergram in the early stages of this work, I found that a sample size of 75 grains was sufficiently representative of a sample of *P. balfouriana.* The measurements of breadth of corpus and of overall breadth were plotted one by one while measuring, and the plots were contoured into an enclosed area. As measurement progresses, the area enlarges until the pollen count reaches 75. From then on the measurement of an additional 20 grains shows that their plots fall within the same area as that of the 75th count. From this result it was decided to make 100 grains the sample size for all measurements (figs. 7, 8).

From calculations of mean values at every 10th count of pollen measurement of *P. monticola,* I found that a sample size of 70 gives a mean value not much different from that obtained from the 103d count. In fact, the mean values of the 80th, 90th, and 100th counts do not differ very much from the mean value of the 70th count (table 7).

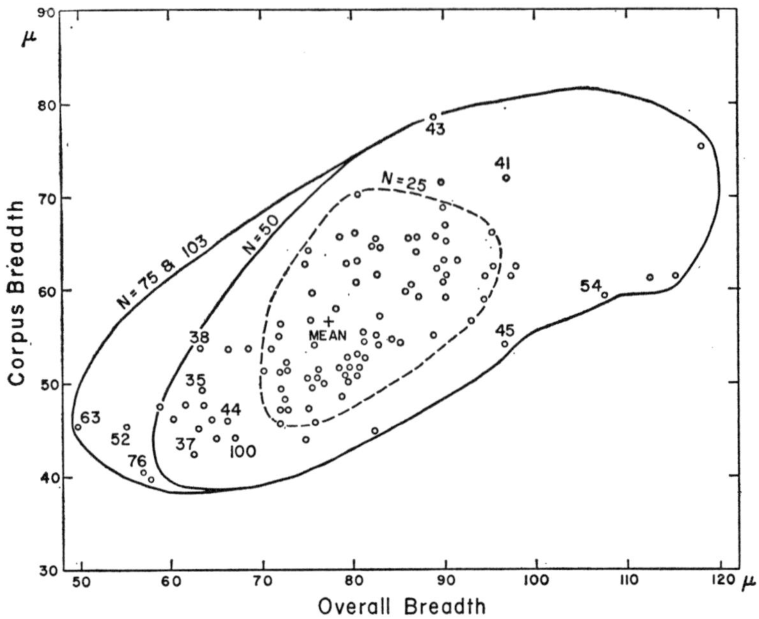

Fig. 7. Scattergram of *Pinus monticola* 2. The scattergram plots measurements of two dimensions. Areas enclosed by plots of $N = 25$, 50, 75, and up to 103 grains, show that area enclosed at end of 75th plot is no different from that at end of 103d plot. Area enclosed at end of 25th plot actually covers three-fourths of all plots. With $N = 50$, 96 per cent of final area is already covered.

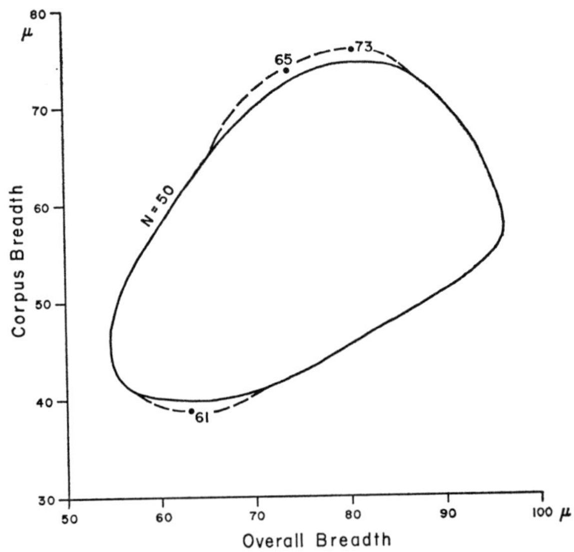

Fig. 8. Scattergram of *Pinus balfouriana.* Area enclosed by solid line at end of 50th plot actually covers 97 per cent of area enclosed by an additional 25 plots (broken line). After 73d plot, no grains fall outside limits indicated by broken line. A sample number (N) of 75 is therefore a good sample size, and a number of 50 constitutes a fair sample.

After studying some 200 samples, I came to the conclusion that the minimum requirement of a sample size varies with the species of pine. For *P. contorta, P. aristata, P. monophylla,* and *P. contorta* var. *bolanderi,* a sample size of 50 grains is often good enough (for *contorta* see table 9). For the other species, 75 to 85 are sufficient. For easier calculations I chose 100. For pollen grains of larger sizes than 70μ, a sample size of 100 is not sufficient. These samples can easily be distinguished from the others, however, merely by their larger mean size of corpus breadth, and a smaller sample size does not represent a real drawback in using the data.

The data in appendix tables 1 and 2 reveal that most samples are drawn with a sample size of 100. Some, which I made earlier, were drawn with a sample size of 50.

Data in which the standard error is unity (= 1.00) are used for purposes other than identification. Additional measurements beyond the original 100 may be needed for many samples to satisfy this condition. In the equation $N = 4SD/SE$, where SD is the standard deviation of mean; SE, the standard error of mean; and N, the sample size, if SE is to be 1.00, then the required sample size N should be four times the value of the variance. These data are not given in appendix tables 1 and 2, for they are not used directly in identification. The required extra measurements beyond 100 can easily be obtained while measuring the pollen grains of the samples.

The Class Interval

Since the raw scores are more conveniently studied by grouping, a decision as to class interval had to be reached early in the course of this study. The grouping could not be too coarse, nor so fine as to hinder manipulation. Identical intervals for all the data made comparison easier.

The grouping must be adjusted to cover the errors of measurement. A grouping interval of 3μ was chosen to cover possible errors in reading the microscopic scale. Since measurement of pine pollen under oil emulsion is very tedious, this 3μ interval enables making measurements without the use of oil lens. A magnification of 650 to 750 times is satisfactory. The average standard error of 236 samples was later found to be 1.063μ, further confirming that a 3μ interval is sufficient to cover the measurement error.

The 3μ interval yields 9–13 classes for the data of corpus measurements. A 5μ interval appears to be too coarse for some of the data. The interval 3μ is approximately half the average standard deviations (= 6.1136μ). It does not satisfy the one-fourth of standard deviation requirement for large samples. For a discussion of class interval, see Wallis (1960, p. 175) and Snedecor (1959, pp. 196–197).

Statistical Data

The individual statistical data of all samples are tabulated in appendix tables 1 and 2. The data for each morphological group, A and B, are in appendix table 3. The data for each species are tabulated in table 8.

From the multitude of reference samples, I have chosen 20—one for each taxon—to illustrate how the various statistics needed were derived (table 9, fig. 9). These 20 samples give a total sample size (N) of 2,000. The average means of

TABLE 8

Statistics and Morphology of *Pinus* Species
(N = 100 in all individual samples used for this computation)

Morphology: Group	Morphology: Pattern	Species	Number of individual samples	$\bar{X}_1$ (in μ)	SD_1 (in μ)	$\bar{X}_2$ (in μ)	SD_2 (in μ)	$\bar{X}_3$ (in μ)	SD_3 (in μ)	Mean of Cos A of samples	SD (in μ)	Ratio $\bar{X}_1/\bar{X}_2$	Ratio $\bar{X}_3/\bar{X}_1$	Thickness of crest (in μ)	Normal thickness of exine (in μ)
A[a]	W[c]	*P. albicaulis* Engelm.	8	59.060	3.164	57.618	1.953	88.936	3.081	0.7561	0.025	1.0	1.5	5	4
		P. flexilis James.	6	55.285	2.430	51.850	4.270	79.405	4.506	0.7198	0.033	1.0	1.4	5	3
		P. monticola Dougl. . .	11	59.010	3.071	54.837	3.255	84.238	5.854	0.7172	0.030	1.0	1.5	5	4–5
		P. monophylla Torr. & Frém.	7	55.190	4.134	44.097	3.501	75.581	5.905	0.6912	0.017	1.2	1.3	5	3
	Y[d]	*P. lambertiana* Dougl.	9	64.633	7.319	65.774	6.020	95.459	9.254	0.7383	0.028	0.9	1.4	5	2–4
		P. aristata Engelm. . .	13	50.846	3.139	48.832	3.795	72.356	5.389	0.7126	0.046	1.0	1.4	11	3
		P. balfouriana Grev. & Balf.	8	53.271	2.856	47.002	3.417	69.768	4.002	0.6452	0.037	1.1	1.3	5	4–5
	W	*P. torreyana* Parry ex Carr.	3	64.900	1.852	58.720	0.874	89.040	0.400	0.6851	0.024	1.1	1.3	. .	3
B[b]	Y	*P. contorta* Dougl. ex Loud.	23	53.744	3.934	44.239	5.022	74.462	6.695	0.6980	0.034	1.2	1.3	. .	3.5–4.5
		P. c. var. *bolanderi* Vasey.	5	50.926	1.876	41.486	2.029	66.836	1.689	0.6494	0.023	1.2	1.3	. .	2
		P. murrayana Grev. & Balf.	29	57.288	4.420	46.428	4.277	78.120	5.780	0.6854	0.038	1.2	1.3	5.5	3.5
		P. coulteri D. Don. . .	10	76.504	5.768	65.694	5.722	107.899	10.794	0.7108	0.038	1.1	1.4	7–16	3.5
		P. jeffreyi Grev. & Balf.	16	76.090	5.007	66.266	4.174	108.501	7.597	0.7212	0.026	1.1	1.4	14.5	6
		P. ponderosa Dougl. ex P. & C. Lawson	33	62.844	4.369	56.816	3.695	90.744	6.850	0.7292	0.032	1.1	1.4	4.0	2.5
	WY[e]	*P. attenuata* Lemmon.	28	62.130	3.486	54.097	4.207	88.894	5.950	0.7240	0.039	1.1	1.4	. .	3
	W	*P. muricata* D. Don. .	13	61.953	5.662	54.409	5.875	87.212	8.465	0.7084	0.025	1.1	1.4	3.5	1
		P. radiata D. Don. . . .	6	62.700	3.719	55.275	2.956	90.188	3.463	0.7266	0.056	1.1	1.4	3–6	2.5–4
		P. remorata Mason. . .	4	59.802	4.472	53.488	1.839	83.922	5.294	0.7034	0.050	1.1	1.4	4	3.5
	WY	*P. sabiniana* Dougl. . .	14	68.439	5.264	58.564	5.262	97.251	6.259	0.7271	0.039	1.1	1.4	7	4.5

[a] Group A. Pollen with sculptured distal membrane.
[b] Group B. Pollen with psilate distal membrane.
[c] W. Beaded pattern on capps.
[d] Y. Graphic pattern on capps.

TABLE 9

MEAN AND SPECTRA OF CORPUS BREADTH OF TWENTY TAXA OF *Pinus*

Species and sample number	$\overline{X}_1$ (in μ)	Pollen frequency by class midpoint (modes in italics)																							
		35	38	41	44	47	50	53	56	59	62	65	68	71	74	77	80	83	86	89	92	95	98	101	104
P. albicaulis 7.......	58.16	..	2	2	8	6	8	*16*	6	6	6	*22*	8	6	0	2	2								
P. flexilis 7.........	58.04	..	..	2	2	6	*14*	10	6	16	*20*	*18*	2	0	4										
P. lambertiana 7.....	72.70	..	..	..	..	2	0	2	2	4	4	*15*	10	6	*21*	6	*12*	*10*	6						
P. monticola 7.......	63.68	..	..	..	..	2	2	8	2	*18*	*18*	*16*	12	12	6	4									
P. aristata 7.........	53.00	..	..	2	2	14	*26*	22	14	8	10	2													
P. balfouriana 8.....	55.52	..	..	2	0	5	10	*33*	15	13	17	5													
P. murrayana 18.....	53.72	..	2	0	*8*	2	18	*36*	14	2	*10*	8													
P. contorta 18.......	56.72	..	..	..	..	..	4	*30*	*32*	12	*16*	6													
P. latifolia 23.......	50.18	..	2	6	8	16	24	*32*	6	4	2														
P. contorta var. *bolanderi* 7(× 2)..	49.88	..	..	8	6	16	*34*	26	8	2															
P. ponderosa 15.....	61.52	..	..	..	1	1	0	9	10	7	*36*	30	1	3	2										
P. jeffreyi 8.........	72.95	..	..	..	..	..	..	..	3	1	1	12	13	16	13	*25*	9	3	2	2					
P. attenuata 20......	58.52	..	..	..	1	2	4	18	17	19	*23*	12	1	3											
P. muricata 13.......	68.60	..	..	..	..	1	0	2	7	6	7	*15*	*15*	*14*	*14*	10	5	4							
P. radiata 7.........	60.20	..	..	..	2	1	0	*19*	13	12	*22*	18	7	4	2										
P. remorata 6.......	63.08	..	..	..	..	..	..	8	8	*20*	12	*24*	20	0	8										
P. monophylla 7.....	55.22	..	..	..	..	2	22	*32*	16	12	8	6	0	2											
P. sabiniana 7.......	65.42	..	..	..	..	..	4	0	*14*	6	*24*	16	6	8	8	4	*10*								
P. torreyana 5.......	66.65	..	..	..	..	..	..	3	2	7	17	*29*	11	13	9	5	4								
P. coulteri 3.........	86.84	..	..	..	..	..	..	..	..	..	..	..	..	5	1	5	11	*16*	10	12	*30*	5	3	1	1
$\overline{\overline{X}}$ = 61.4945; total spectrum		..	6	22	38	76	170	*306*	195	175	*253*	*254*	106	92	88	61	53	33	18	14	*30*	5	3	1	1

corpus, saccus, and overall sizes are $\overline{\overline{X}}_1 = 61.49\mu$, $\overline{\overline{X}}_2 = 53.90\mu$, and $\overline{\overline{X}}_3 = 86.25\mu$, respectively. The standard deviation of $\overline{\overline{X}}_1$ increases when sample size is increased because of mixing of the universes. The standard error, calculated from $SD/N^{1/2}$, diminishes from 1.74μ, highest among the samples, to 0.23μ. The confidence interval is taken as the equivalent of approximately twice the standard error: that is, for corpus, 0.46 $(1.96\ SD/N^{1/2})$.

RATIO INDEX

Ratios between means are obtained by dividing the means of corpus breadth, of saccus breadth, and of the overall breadth by the mean of saccus breadth, giving

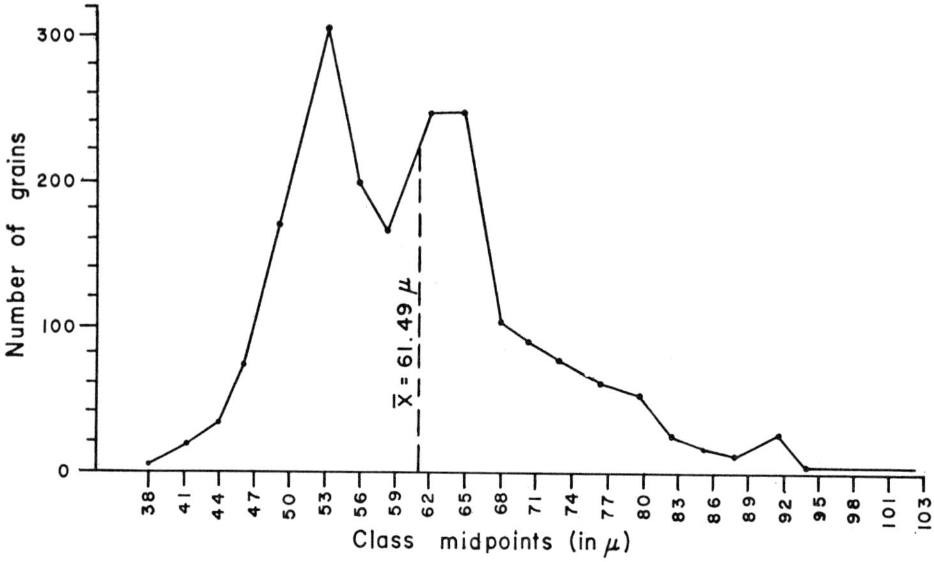

Fig. 9. Frequency of sizes of corpus breadth from twenty pollen samples belonging to twenty species of *Pinus,* each numbering 100 grains. The frequency curve shows two prominent modes and a minor mode at class 92. Data: $\overline{X} = 61.49\mu$; $SD = 10.725$; $2SD/N^{1/2} = 0.46$; coefficient of variance, $C = 17.45$ per cent; cos $A = 7062$.

R_1, R_2, and R_3, respectively. Divide the mean of overall size by the mean of the corpus breadth to obtain R_4. In the case of the 20 samples studied above, $R_1 = 1.140$, $R_2 = 1.000$, $R_3 = 1.600$, and $R_4 = 1.403$. Disregard the decimal point and take the first two digits of R_1 and R_4 (i.e., 1.1 and 1.4) and combine them in that order, giving the ratio index of 1114.

If we consider also the possible range of the mean $(\overline{X}_1 \pm 2SE)$ and obtain the ratios from these extreme values, they may turn out to be $R_1 = 1.2$ and $R_4 = 1.3$. Hence the indexes of this mixed sample of 20 pines may be four in number: 1114, 1113, 1213, 1214. For our purpose, the one ratio obtained by using only the mean value is sufficient; the other ratios obtained by mean $\pm$ $2SE$ are not necessary.

The index of 1114 is the mean index for *all* California pines. Actually, pollen grains of some species differ widely in their indexes. In the *P. contorta* group (*P. contorta, P. murrayana,* and their varieties) indexes of 1213 or 1214 are common, and two samples have 1412. The indexes for the white pines are apt to be 914, 915,

1013, 1014, or even 1015 and 1016. This sort of index provides one of the clues for identification of unknown species.

COS A INDEX

Let three variables x_0, y_0, and z_0 be plotted on three axes XX', YY' and ZZ' (fig. 10). If these three variables equal one another, the lines connecting them form an

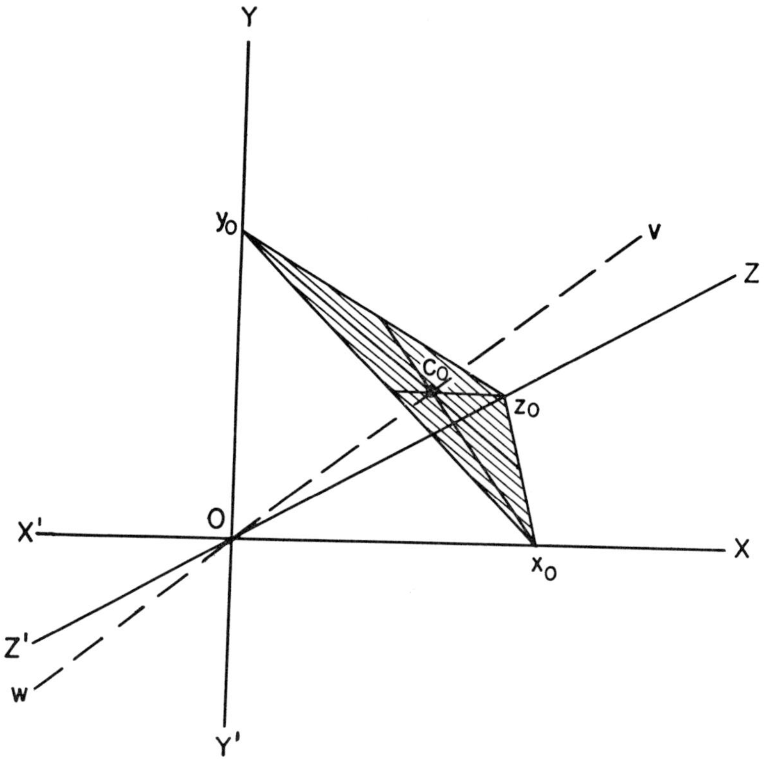

Fig. 10. Three-dimensional plot for variables x_0, y_0, z_0. The three axes XX', YY', ZZ' are perpendicular to each other. Center of triangle, C_0, lies on a straight line of view, *vw*, which passes throught the origin, 0.

equalateral triangle. Center C_0 of this triangle will then be the focus of three intersecting lines, each of which subdivides equally the opposite side of the angle where the line originates. This is shown in fig. 10 in oblique view. The center C_0 is on a line *vw* which passes through the origin. The same data are also shown in figure 11, where the view is at 45° and the center C_0 coincides with the origin.

Since x, y, and z are variables and do not necessarily equal one another, the triangle formed by joining them is not always equilateral. Triangles formed by sets of values x_1, y_1, z_1 and x_2, y_2, z_2 are shown in figure 11; their centers are C_1 and C_2, respectively.

If x, y, and z represent respectively the mean dimensions of corpus, overall breadth, and saccus of pollen of a pine species, the triangle or the center of the triangle can be used as an indicator for that species. Hence, each of the loci of centers 1 and 2 shown in figure 11 represents a distinct species of pine, and the loci are far apart. The shapes of the triangles also differ.

We know that the mean value deviates within a range determined by the confidence interval. Allowing for such deviations, each of the triangles will, in fact,

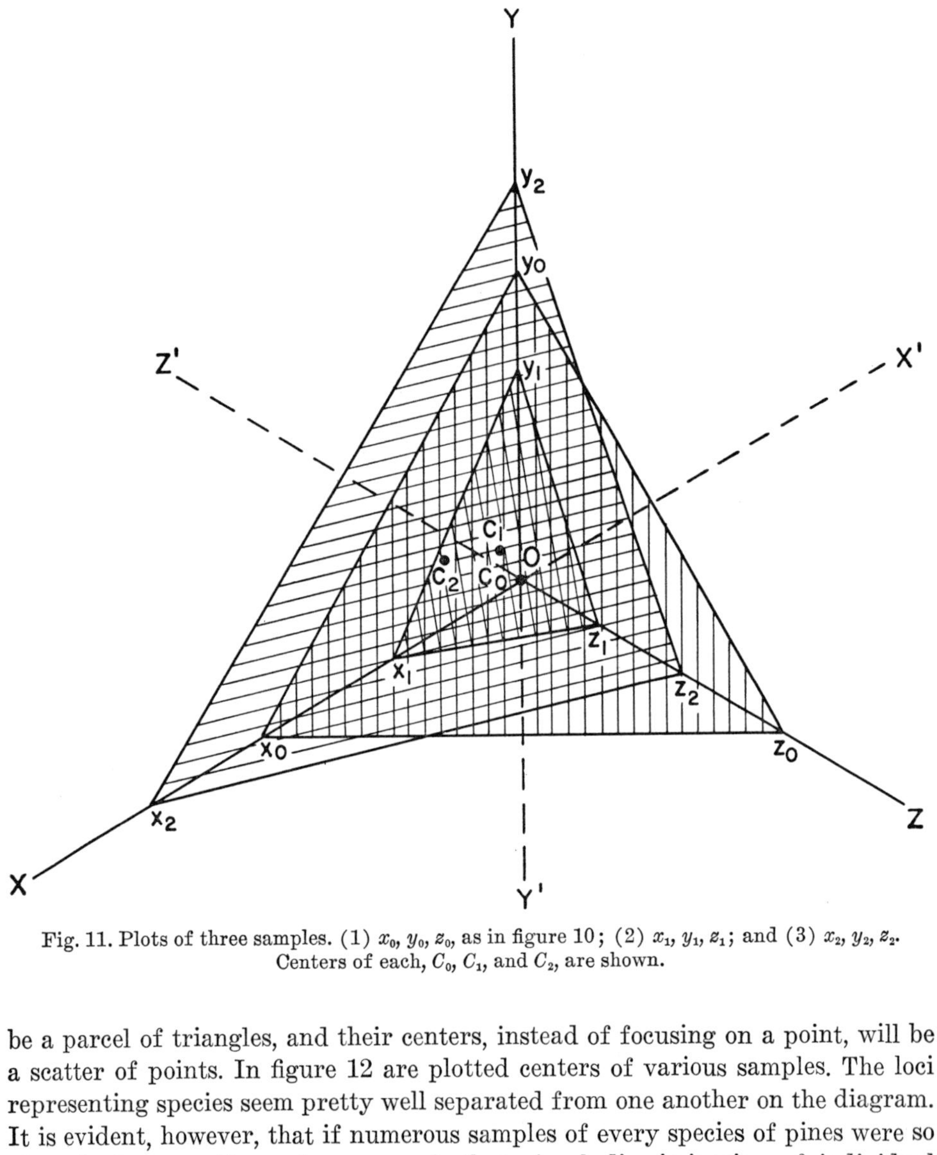

Fig. 11. Plots of three samples. (1) x_0, y_0, z_0, as in figure 10; (2) x_1, y_1, z_1; and (3) x_2, y_2, z_2. Centers of each, C_0, C_1, and C_2, are shown.

be a parcel of triangles, and their centers, instead of focusing on a point, will be a scatter of points. In figure 12 are plotted centers of various samples. The loci representing species seem pretty well separated from one another on the diagram. It is evident, however, that if numerous samples of every species of pines were so plotted, they would overlap so much that visual discrimination of individual samples, or of species from one another, would not be possible. For this reason a numerical index was sought as a substitute. This is the cos A index.

The cos A index is found by the law of cosines. Let the three sides of a triangle (x, y, z in fig. 12) be the dimensions of corpus, overall size, and saccus, respectively.

The angle between the lines representing overall and saccus dimensions and facing that of corpus is the angle *A*. Expressed as a trigonometric function,

$$\cos A = \frac{b^2 + c^2 - a^2}{2bc}$$

where $a = x_0$ is the dimension of corpus, $b = y_0$ is the dimension of overall breadth, and $c = z_0$ is the dimension of saccus. The formula can be simplified by dividing the values *a*, *b*, and *c* by *c*, since in our pine measurements *c* is usually the shortest. Substitute the ratio of $R_1 = a/c$, $R_3 = b/c$, and $R_2 = c/c$ (R_4 is b/a) into the equation; it will then be reduced to

$$\cos A = \frac{1 + R^2{}_3 - R^2{}_1}{2R_3}.$$

The natural cosine value is used as the index. There is no need to transform the value into degrees, minutes, and seconds. For the example shown in figure 12, cos A is 0.6000. Ignore the decimal and use 6000 as the index. For the example of 20 pines given, the mean cos A index is 7062.

There remains one point that needs explanation. We have constructed the triangles by marking off the dimensions of corpus, saccus, and overall size on the axes (figs. 10, 11). The lines that join the intercepts, and thus form the triangle, are the dimensions we measured. In figure 12, however, we made the length of the sides of the triangle represent the ratios. Given the same data, the calculated cos A from a triangle constructed by dimensions such as those in figure 11 will be different from those constructed by ratios. But this does not affect us in any way as long as we arbitrarily adhere to one or the other method of obtaining the triangles. By using ratios as the sides of triangle, we eliminate considerable calculation and plotting. The ratios are therefore the preferable values for obtaining the *cos A* index.

The cos A indexes of each species studied do show some tendencies of specific value. For instance, the indexes of *P. balfouriana* are always below 7000, whereas those of *P. albicaulis* are always above 7000. This fact enabled me to spot an error of specimen identification in the herbarium, which was verified by Critchfield. Indexes between 6100 and 6300 seem confined to the *P. contorta* group. Nevertheless, the index cannot be relied on solely for specific identification because it is also distributed amid a range within which many species have closely similar indexes. Experience shows that the index, when used together with other criteria, usually provides the final clue to the correct identification with a high degree of consistency.

SPECTRUM AND MODAL DISTRIBUTION

A spectrum consists of pollen frequencies of all classes, in order of class magnitude. Our example of a spectrum formed by 2,000 grains, from 20 species, is shown at the bottom of table 9 in the total spectrum.

There have been many attempts to render such figures visually. The frequency curve, showing the distribution of sizes of pollen, is one of the most popular. But a frequency curve can only show one measurement at a time: either the corpus, the saccus, or any other measurement. Identification of unknowns from frequency curves is based on two assumptions: (1) that for every species of pine, the distribution is unimodal, and (2) that the number of modes of a frequency distribution indicates the number of species forming the spectrum. Neither of these assumptions is completely justified. As we shall see, the unimodal and bimodal distributions are equally abundant among my references, and the example of 2,000

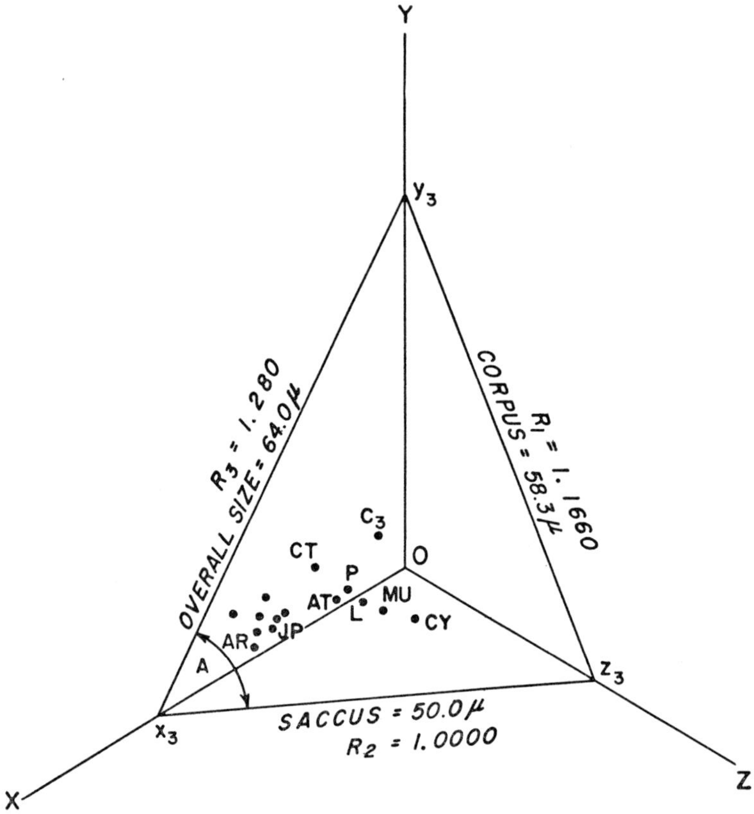

Fig. 12. Loci of centers of three-dimensional plots. Three-dimensional plots of loci of centers representing different taxa and a hypothetical sample: C_3 (x_3, y_3, z_3). The individual samples can be plotted from their mean values or from their ratios (R_1, R_2, R_3). Cos A is 6000.

Abbreviations: *AR*, *Pinus aristata; AT*, *P. attenuata; CT*, *P. contorta; CY*, *P. murrayana; JP*, *P. jeffreyi* × *ponderosa; L*, *P. lambertiana; MU*, *P. muricata; P*, *P. ponderosa; A*, angle to be expressed as cos A.

grains from 20 pine species has only three modes (fig. 9, table 10). The number of modes depends on the sample size and the class interval of the spectrum. It would not be valid to match the number of species to the number of modes in the present data. It is not even possible to identify the modes with the majority of modal distribution of the 20 individuals. For instance, the mode at class 53, which has a total of 306 grains, is the mode of 7 of the 20 spectra. The mode at class 65 occurs in only 5 of the 20 spectra, and the minor mode at class 92 belongs exclusively to the spectrum of *P. coulteri*. Actually there are spectra of 10 species having a mode at class 62, yet their sum of frequencies is comparable only to that of class 65. It has been pointed out to me that frequency curves and modal number may work very nicely in a geographical area where there are only a few existing species, say three. I would not agree with this statement if the data are based on a sample of 100 grains and with a 3μ class interval. An educated guess at identifying one of only three species has a high probability of being correct without even consulting the modes.

The modal distribution is but another measure of the central tendency. It may not be necessary if the mean values are sufficient for identification of a species. It does play an important role in the fitting of the spectrum of fossil samples discussed later.

A mode is the class where the highest frequency occurs in a histogram. A frequency distribution theoretically has only one mode. Usage has led to the application of the terms "major mode" to the class with the highest frequency, and "minor modes" to those classes with lower frequencies in a spectrum, provided each of them stands out as a peak on the curve. I have further limited a mode to such a "peak" whose frequency is more, by 3 per cent, than its adjacent classes. The major mode in the example of 20 pines is at class 65, but the frequency of class 62 differs from it only by one. It is therefore a co-mode. The modes at classes 53 and 92 are minor modes. Not infrequently the end class of a spectrum has a higher frequency than its immediate neighbor. Unless their difference is more than 5 per cent, it is not considered a mode. Use the class midpoints to describe the modes. In writing always put the major mode first, followed in descending order of frequency by the minor modes; thus the modes of the example (fig. 9, table 10) are described as 65–62, 53, 92.

The number of modes in a spectrum among 313 samples (table 10) shows approximately equal prevalence of unimodal (49.6 per cent) and bimodal (42.0 per cent) distributions. The trimodal (8.0 per cent) and the multimodal (0.4 per cent) distributions are rare.

Table 10 shows the modal distribution in each species of pine according to frequency of occurrence. Only *P. lambertiana* has a tendency to be dominantly trimodal.

Among the modal patterns, the unimodals are concentrated in class 53 among the Group A pines and in classes 65 and 53 among the Group B pines. Each individual species has to be considered separately because all species do not have the same number of samples. The total frequency of class 65 has certainly been exaggerated because we have more samples of *P. ponderosa* and *P. attenuata* than of others, and both of them have dominance in mode 65.

Shifting of modes among samples or lots is common. In the process of assembling the data I noticed that in some samples the position of the major mode is very constant, whereas in others it changes from lot to lot. Reviewing all the data, I found that there is a relation between modes and a combination of certain modes and species. For instance, among the pollen spectra of 35 samples of *P. ponderosa* (see appendix table 2), the 62 class turns up 12 times as a dominant mode, and the 65 class 16 times. On the other hand, among the 22 spectra of *P. jeffreyi*, the major mode is at class 62 only 2 times, at class 65, 3 times, but at class 74, 6 times,

TABLE 10

MODAL DISTRIBUTION AND DOMINANT MODES AMONG *Pinus* SPECIES

Group	Species	Frequency of modal pattern					Dominant modes			Dominant modal pattern
		Unimodal	Bimodal	Trimodal	Multimodal	Total	Unimodal	Bimodal	Trimodal	
A	*P. albicaulis*	2	6	1	..	9	53	53, 62	59, 65, 53–50	bimodal
	P. flexilis	1	6	..	..	7	47	50, 62	....	bimodal
	P. monticola	5	3	4	..	12	53	62, 53	62, 53, 59, 41–44	unimodal
	P. monophylla	5	4	1	..	10	53	53	62, 53, 44	unimodal and bimodal
	P. lambertiana	2	2	4	1	9	53, 59	65	65, 74, 80, 56	trimodal
	P. aristata	10	4	1	..	15	53	41	53, 59, 44	unimodal
	P. balfouriana	6	2	2	..	10	53	53	56–53, 50, 44	unimodal
	P. torreyana	3	1	..	..	4	65	65, 71	....	unimodal
	Subtotal	34	28	13	1	76	53	53, 62	53, 62–65, 59, 44	unimodal
	Percentage	44.7	36.8	17.1	1.4	100.0				
B	*P. contorta*	10	12	..	..	22	53	53, 62	....	unimodal and bimodal
	P. contorta var. *bolanderi*	5	2	..	..	7	53	50	....	unimodal
	P. murrayana	23	15	..	..	38	53	53, 62	....	unimodal
	P. jeffreyi	12	15	3	..	30	74	74	65, 53	unimodal and bimodal
	P. ponderosa	29	18	3	1	51	65	65, 53	53, 65–68	unimodal
	P. coulteri	2	6	1	..	9	77	80	80, 71–68, 86	bimodal
	P. attenuata	21	14	1	..	36	65, 62	62, 65, 53	62, 71, 53	unimodal
	P. muricata	9	5	1	..	15	68	53, 62	53, 62, 68	unimodal
	P. radiata	4	3	1	..	8	62, 59	62	71, 62–59, 77	unimodal and bimodal
	P. remorata	1	3	..	..	4	53	62	....	bimodal
	P. sabiniana	5	10	2	..	17	65	74, 65, 77	65, 71, 56	bimodal
	Subtotal	121	103	12	1	237	65	53, 62	53, 65, 62, 71	unimodal
	Percentage	51.0	43.6	5.0	0.4	100.0				
Grand total		155	131	25	2	313				
Percentage		49.6	42.0	8.0	0.4	100.0				

at class 77, 5 times, and at class 80, 4 times. The last three modes never appear on the spectra of *P. ponderosa.*

Among the 7 spectra of *Pinus contorta* var. *bolanderi* (appendix table 2), classes 50 and 53 each serve as the dominant mode in three cases, whereas in *P. contorta* the dominant mode at class 53 occurs in 13 of 19 spectra and a dominant co-mode of 50–53 occurs in only one spectrum. The closely related *P. murrayana* has mode dominance in class 53 (16 out of 25), but it also occurs 9 times out of 25 in class 62, which never occurs in the spectra of *P. contorta* var. *bolanderi* and only once in the spectra of *P. contorta.* In the interpretation of spectra of unknown single species, a study of modal distribution of these reference spectra aids in identification in much the same way that a table of classification provides alternate choices of subgroups of species. Figure 13 shows the modal distribution in digrammatic form. When searching for identities, first classify the spectrum with regard to mode type; uni-, bi-, or trimodal; then find the species that match the modes.

COEFFICIENT OF VARIANCE

The quotient found by dividing the standard deviation of mean by the mean is the coefficient of variance (C). It is usually expressed as a percentage. The coefficient commonly employed for checking the consistency of experimental results is here used to indicate the heterogeneity of samples. I noticed, while studying data of mixed samples and fossil samples, that the C of such a sample is often higher than the C of most of the pure samples. For instance, my example of a mixture of 20 pines ($N = 2{,}000$) gave a C of 17.5 per cent, a value exceeded only by one pure sample out of a total of 276 samples (appendix table 2, *P. coulteri* 1). Of the 276 samples, the average is only 9.7 per cent ± a $2SE$ of 0.78 per cent.

From the original equation for C (for single species $C = SD/\overline{X}$ for mixtures $C_m = SD_m/\overline{X}_m$) a change in C is obviously directly proportional to the change in SD and inversely proportional to the change in $\overline{X}$. It is the value of SD which is more decisive, since the mean of a mixture ($\overline{X}_m$), say, of two components, may assume different values from their original individual means, but will never be larger than the higher nor smaller than the lower mean of the original components. If these two components are identical, C_m will always be smaller, because the SD_m will always be smaller than the original (SD). If the two components differ only slightly in their mean values, say that the difference is statistically insignificant, C_m is still largely dependent on SD_m.

Standard deviation of mean of a sample is affected by (1) the distribution of frequencies in the spectrum, and (2) the number of frequencies in the sample or the sample size. In mixing of a number of components, N is increased as many fold as there are components. Assuming the fixed sample size for every component is 100, the mixing of two makes N 200, of five makes it 500. The increase of N depresses the SD only if the frequency distributions of the two samples are alike. If an increase in N results in an increase in SD, the increase is not owing to the effect of larger N but to the distribution of frequencies, which must vary in such a way that the sum of squares is greatly increased.

The frequency distribution is expressed in the spectrum of a sample. The properties easily found in a spectrum are the length—in other words, the number of

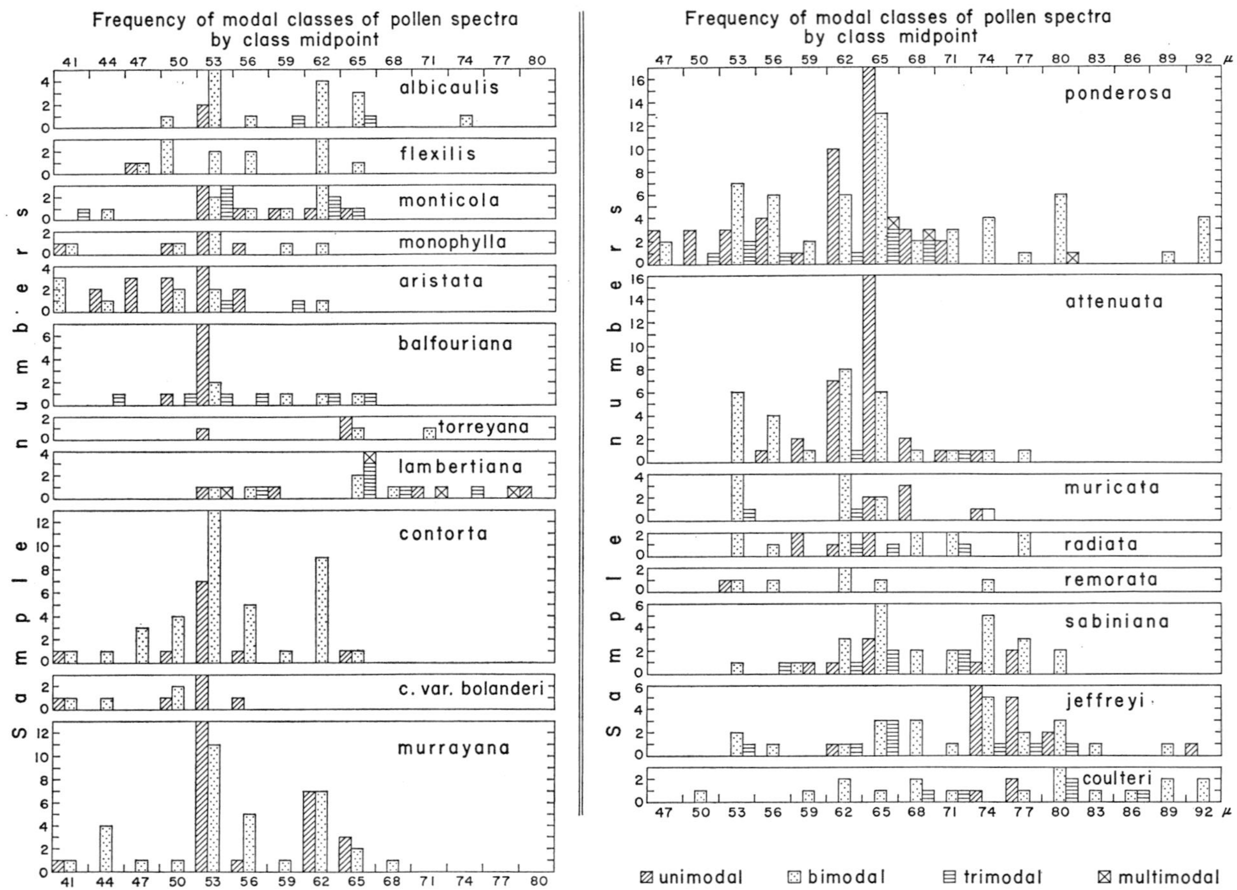

Fig. 13. Histograms showing frequency of modal classes of pollen spectra (breadth of corpus) of *Pinus* species in California.

classes in grouped data—and the range. By mixing two spectra of equal N, of nonsignificantly different $\overline{X}$, and of identical range, the increase in N does not significantly alter the standard deviation of mean. The C_m that results from mixing does not change very much and usually is somewhat depressed. On the other hand, if two components of different class number and of different range, but with nonsignificantly different means, are mixed, the standard deviation may be significantly elevated, resulting in a higher C_m.

TABLE 11

RANGE AND MEAN OF CLASS NUMBER AND OF COEFFICIENT OF VARIANCE

Species	Number of samples	Class number		C			Class-C index
		Range	$\overline{X}$	Range	ΣC/N	$SD_T/\overline{X}_T$	
P. albicaulis	9	7–15	10.2	7–15	11.5	5.3	10–12
P. flexilis	6	9–12	10.8	10–13	11.8	4.4	11–12
P. lambertiana	9	9–16	12.7	9–15	11.9	11.3	13–12
P. monticola	11	9–12	10.2	8–14	10.4	5.2	10–10
P. aristata	13	7–14	9.7	8–14	12.1	6.1	10–12
P. balfouriana	8	8–14	10.5	9–12	10.6	5.3	10–11
P. monophylla	7	8–11	9.5	8–11	10.7	7.5	10–11
P. torreyana	4	9–14	11.0	7–10	8.8	2.8	11–9
P. contorta	22	6–12	8.2	6–12	8.8	7.3	8–9
P. c. var. *bolanderi*	4	7–12	9.3	7–12	9.8	3.7	9–10
P. murrayana	32	6–14	9.4	7–13	9.8	7.7	9–10
P. jeffreyi	24	9–20	13.5	7–17	10.7	6.7	14–11
P. ponderosa	44	7–16	11.5	5–14	9.7	6.9	12–10
P. coulteri	10	8–16	15.0	6–20	12.2	7.5	15–12
P. attenuata	32	8–16	11.0	7–13	9.4	5.6	11–9
P. muricata	14	8–13	10.8	7–15	9.7	9.1	11–10
P. radiata	6	10–16	12.3	10–16	10.2	5.9	12–10
P. remorata	4	8–9	8.3	6–9	7.6	7.4	8–8
P. sabiniana	14	9–16	11.3	8–13	10.0	7.7	11–10

SYMBOLS: ΣC Sum of coefficients of variance of individual samples within one taxon.
SD_T Standard deviation of taxon.

In order to understand the conditions under which C may be useful in deciding whether a sample is homogeneous or heterogeneous (i.e., composed of more than one universe), I made some tests. The tests are designed in such a way as to show the effects on C and C_m by alternately setting either the class number, or the range, or the mean, or the SD individually, or in combination, as constants.

Test I (tables 12,13*a*,*b*).—Class number and range of the original spectrum are identical and are maintained as constants in the mixed samples.

Seven samples of equal class number are chosen from the references (appendix table 1) for testing. Each of these is mixed with one of the others and for each of the mixed samples the $\overline{X}_m$, the SD_m, the C_m, and the proportion between means ($\overline{X}_{ma}$: $\overline{X}_{mb}$) are calculated. These values are tabulated in tables 13*a*, *b*. To read the tables, take, for example *P. balfouriana* 4 (code 1 in table 13*a*) and mix it with *P. murrayana* 30 (code 7); their average mean, $\overline{\overline{X}}$, is $\frac{1}{2}$ $(51.32+57.68)\mu=54.50\mu$. This value appears in table 13*a*, column 7, row 1. The proportion of the means is 57.68/

51.32 = 1.12, which appears in table 13*b*, column 7, row 1. Between two means, a proportion higher than 1.05 signifies that the two means are statistically different. All the means that are significantly different are marked off from the nonsignificant ones by a heavy zigzag line in the table. Since the proportion between the two means is significant, the mixing of the two may be expected to give a spectrum whose SD_m is higher than the original individual *SD* of the samples. This is true in this case. The SD_m is found to be 6.38μ (table 13*a*, column 1, row 7). The original *SD* for *P. balfouriana* 4 is 5.48μ, and for *P. murrayana* 30 is 5.63μ. Since *C* varies according to *SD* and $\overline{X}$, the C_m of the mixed sample is probably also high. It is found to be 11.7 per cent (table 13*b*, column 1, row 7), whereas the original *C* values are 10.7 and 9.8 per cent, respectively.

TABLE 12

SPECTRA OF SEVEN INDIVIDUAL *Pinus* POLLEN SAMPLES OF EQUAL CLASS NUMBER (10) AS IN TEST I

Species and sample number	*N*	Pollen frequency by class midpoint									
		41	44	47	50	53	56	59	62	65	68
P. balfouriana 4	100	7	7	12	25	30	6	5	7	0	1
P. flexilis 2N	100	5	2	22	20	15	25	4	5	1	1
P. aristata 16	101	1	0	6	18	27	29	10	5	4	1
P. monophylla 6	100	2	8	2	12	22	12	10	26	4	2
P. muricata 10	100	3	2	2	6	32	20	9	20	5	1
P. ponderosa 36	100	2	2	2	8	14	40	12	10	4	6
P. murrayana 30	100	2	0	2	8	22	10	18	24	12	2
Total sample	701	22	21	48	97	162	142	68	97	30	14

NOTE: $\overline{X}_1$ = 54.78μ; SD_m = 5.8933; C = 10.758.

The results of such comparisons for these seven samples are as follows:

1. Mixing two samples with significantly different means results in a C_m value larger than any one of the components.

2. Mixing two samples with means not significantly different from each other does not give a C_m larger than that of each of the individual samples.

3. Mixing two samples of equal variance may result in a slightly elevated C_m, though their means may not be significantly different (*P. muricata* 10 and *P. murrayana* 30 in table 13*a, b*, codes 5, 7).

4. Mixing samples with great difference in *C* may not give an elevated C_m after mixing, though their means may be significantly different (*P. aristata* 16 and *P. balfouriana* 4 in table 13*a, b*, codes 3, 1).

TABLE 13*a*

COMPARISON BETWEEN STANDARD DEVIATIONS OF MIXTURES OF TWO COMPONENTS AND OF THE INDIVIDUALS BEFORE MIXING AS IN TEST I

Code number	Species and sample number	$\overline{X}$ (in μ)	SD (in μ)	Code number of species and sample: 1	2	3	4	5	6	7
				Average mean, $\overline{\overline{X}}_m$ (in μ)						
1	*P. balfouriana* 4	51.32	5.48	....	51.66	52.92	53.48	53.58	53.87	54.50
2	*P. flexilis* 2N	52.01	5.41	5.44	.....	53.27	53.82	53.93	54.22	54.84
3	*P. aristata* 16	54.52	4.40	5.29	5.16	.....	55.08	55.17	55.47	56.09
4	*P. monophylla* 6	55.64	6.40	6.31	6.47	5.58	.....	55.74	56.03	56.66
5	*P. muricata* 10	55.85	5.48	5.92	5.76	5.10	5.93	.....	56.14	56.76
6	*P. ponderosa* 36	56.42	5.49	6.03	5.80	5.12	5.93	5.46	.....	57.05
7	*P. murrayana* 30	57.68	5.63	6.38	6.18	5.35	6.03	5.60	5.55	.....
	Average	54.78		SD_m (in μ)						

TABLE 13*b*

COMPARISON OF THE COEFFICIENTS OF VARIANCE OF MIXTURES OF TWO COMPONENTS WITH THEIR ORIGINAL VALUES AND THE PROPORTION BETWEEN THEIR MEANS AS IN TEST I

Code number	Species and sample number	$\overline{X}$ (in μ)	C (in percent)	Code number of species and sample: 1	2	3	4	5	6	7
				Proportion between means						
1	*P. balfouriana* 4	51.32	10.7	....	1.01	1.06	1.08	1.09	1.10	1.12
2	*P. flexilis* 2N	52.01	10.4	10.5	....	1.05	1.07	1.07	1.08	1.11
3	*P. aristata* 16	54.52	8.0	10.0	9.7	....	1.02	1.02	1.03	1.06
4	*P. monophylla* 6	55.64	11.5	11.8	12.0	10.1	....	1.00	1.01	1.04
5	*P. muricata* 10	55.85	9.8	11.0	10.7	9.2	10.6	....	1.01	1.03
6	*P. ponderosa* 36	56.42	9.7	11.2	10.7	9.2	10.6	9.7	....	1.02
7	*P. murrayana* 30	57.68	9.8	11.7	11.3	9.5	10.6	9.8	9.7	....
	Average	54.78		C_m (in percent)						

Test II (table 14).—A hypothetical spectrum is made up of frequencies 1, 2, 3, 2, 1; $N = 9$; class number is 5, the mean of sample is 3. This number is equivalent to the midpoint of class 3 and is also the mode. Mixtures are formed by shifting one spectrum one class up or down from the preceding mixture, and then adding the frequencies class by class. Any mixing of multiples of the same spectrum will result only in decreasing SD_m and C_m, while the $\overline{\overline{X}}_m$ remains exactly the same as $\overline{X}$ of the individuals. By shifting one of the two duplicated spectra at steps of one class each time, we obtain a different spectrum of mixtures with increasing class number, and varying $\overline{\overline{X}}_m$ and SD_m. Note that we are mixing two spectra of the same sample size and the same variance. The shifting of the spectrum of one, up or down the class scale, with a constant sample size and variance produces changes of the C also. The mean, $\overline{\overline{X}}$, can be kept constant if we shift the spectra simultaneously at equal magnitude, but in opposite directions.

The results follow:

1. Mixing, and not duplication, always increases the number of classes whether or not the mean ($\overline{X}_m$) is constant or changes in value. The values of SD_m and C_m will be higher beyond a certain degree of mixing.

TABLE 14

STATISTICS OF ORIGINAL AND MIXED *Pinus* SPECIES AS IN TEST II

Sample	N	Class number	$\overline{X}$ and $\overline{X}_m$ (in μ)	C and C_m (in percent)	SD and SD_m (in μ)	C_m with constant $\overline{X}_m$ (in percent)	C_m with decreasing $\overline{X}_m$ (in percent)
Original.	9	5	3	40.8	1.225	....	....
Mixture							
1.	18	5	3	39.6	1.188	39.6	39.6
2.	18	6	3.5	37.0	1.295	....	....
3.	18	7	4.0	39.3	1.572	52.4	78.6
4.	18	8	4.5	43.3	1.948	....	....
5.	18	9	5.0	47.5	2.376	79.2	237.6

NOTE: The original spectrum consists of five classes, with frequencies 1, 2, 3, 2, 1, respectively. Mixture 1 is the spectrum plus a duplicate of itself. Mixtures 2–5 are formed by shifting toward the higher end one spectrum one class up from the preceding mixture.

2. There is an initial decrease in SD_m when range is not changed and N is increased to $2N$ (table 14, mixture 1).

3. The C_m decreases as classes increase until the spectrum is 3 classes longer than the originals. The C_m increases thereafter (table 14, mixtures 4, 5).

4. If a constant mean value ($\overline{X}_m$) is maintained, or if it is decreased by mixing two identical spectra with different means, the increase of C_m is sharp (table 14, last column).

Test III (table 15).—Two original spectra, A and B, have the same class number but different sample size and variance. The mixed samples have an increasing mean ($\overline{X}_m$).

The results follow:

1. The initial decrease in SD_m continues until the class number is 4 classes longer than the original spectra, whereas C_m decreases as classes increase until the spectrum is 3 classes longer than the originals, identical to the finding in Test II; result no. 3, above.

2. With a constant $\overline{X}_m$, the increase of C_m is even faster than increasing means.

Test IV (tables 16, 17).—Two spectra, one of *P. torreyana* 6T and the other of *P. attenuata* 28, are mixed. The spectra are different in mean and class number,

TABLE 15

STATISTICS OF ORIGINAL AND MIXED *Pinus* SPECIES AS IN TEST III

Sample	N	Class number	$\overline{X}$ and $\overline{X}_m$ (in μ)	C and C_m (in percent)	SD and SD_m (in μ)	C_m if with constant $\overline{X}_m$
Original						
A...............	13	5	3	36.0	0.599	....
B...............	9	5	3	40.8	0.817	....
Mixtures of *A* and *B*						
1...............	22	5	3	37.1	0.474	37.1
2...............	22	6	3.59	34.0	0.521	40.7
3...............	22	7	4.18	35.9	0.640	50.0
4...............	22	8	4.77	44.8	0.780	62.5
5...............	22	9	5.33	55.0	0.981	76.6

TABLE 16

STATISTICS OF ORIGINAL AND MIXED *Pinus* SPECIES AS IN TESTS IV AND V

Species and sample number	N	Class number	$\overline{X}$ (in μ)	C (in percent)	SD (in μ)	
P. torreyana 6T	100	14 (11)	65.09	8.68	5.65	.....
P. attenuata 28.	100	10	62.78	8.94	5.61	.....
Mixture	N	Class number	$\overline{X}_m$	C_m	SD_m	SE (in μ)
Test IV						
1...........	200	14	63.935	8.90	5.68	0.803
2...........	200	16	63.935	10.96	7.01	0.991
3...........	200	18	63.935	14.19	9.07	1.283
4...........	200	20	63.935	18.14	11.60	1.641
5...........	200	22	63.935	22.39	14.31	2.024
6...........	200	24	63.935	26.78	17.12	2.421
Test V						
1...........	150	14	64.30	8.81	5.67	0.926
2...........	150	16	64.30	10.43	6.81	1.113

but have approximately the same SD and C and a fixed sample size, N. The results are that, while maintaining the $\overline{X}_m$ at a constant value, but increasing the class number by 2 while mixing, the values of both SD_m and C_m increase without any setbacks. Note that the difference between the means of the two species is not significant.

Test V (tables 16, 17).—Mixing the same two spectra as in Test IV, but in different proportions—*P. torreyana* 6T, 100 grains, and *P. attenuata* 28, 50 grains (table 16, bottom)—gives the same results. The C is indicative when the mixed spectrum is increased by 2 classes.

TABLE 17

SPECTRA OF ORIGINAL AND MIXED *Pinus* SPECIES AS IN TESTS IV AND V

(In μ)

Original *Pinus* species and sample number	$\overline{X}_1$	Pollen frequency by class midpoint																							
		35	38	41	44	47	50	53	56	59	62	65	68	71	74	77	80	83	86	89	92	95	98	101	104
P. torreyana 6T	65.09						1	2	4	6	25	31	11	14	3	1	1	0	0	1					
P. attenuata 28	62.78						4	6	6	14	14	40	10	0	4	2									
Mixture	$\overline{\overline{X}}_1$																								
Test IV 1	63.935						5	8	10	20	39	71	21	14	7	3	1	0	0	1					
2	63.935					4	6	7	16	18	46	35	31	15	18	3	1	1	0	0	1				
3	63.935				4	6	6	14	15	42	14	6	29	33	11	14	3	1	1	0	0	1			
4	63.935			4	6	6	14	14	40	11	2	8	8	25	31	11	14	3	1	1	0	0	1		
5	63.935		4	6	6	14	14	40	10	0	5	4	4	6	25	31	11	14	3	1	1	0	0	1	
6	63.935	4	6	6	14	14	40	10	0	4	2	1	2	4	6	25	31	11	14	3	1	1	0	0	1
Test V 1	64.300						3	5	7	13	32	51	16	14	5	2	1	0	0	1					
2	64.300					2	3	4	9	11	26	30	31	13	15	3	1	1	0	0	1				

TABLE 18

Spectra of Original *Pinus* Species and of Mixtures of Two as in Test VI

Code	Species and sample number	$\overline{X}_1$ (in μ)	C (in percent)	Pollen frequency by class midpoint																								
				32	35	38	41	44	47	50	53	56	59	62	65	68	71	74	77	80	83	86	89	92	95	98	101	104
a	*P. contorta* 5	61.91	9.3						1	1	10	12	11	23	28	5	5	2	2									
b	*P. murrayana* 15	56.60	8.0						2	8	28	26	10	18	8													
c	*P. muricata* 7	56.51	8.2				4	1	8	17	23	6	6	14	6	9	4	0	2									
d	*P. balfouriana* 8	55.52	5.2				2	0	5	10	33	15	13	17	5													
e	*P. aristata* 1	45.05	5.5	2	6	9	15	19	18	18	12	1																
f	*P. lambertiana* 7	72.70	8.8						2	0	2	2	4	4	15	10	6	21	6	12	10	6						
g	*P. sabiniana* 4	77.87	6.4									1	1	0	2	2	8	18	26	18	13	3	4	3	1			
h	*P. jeffreyi* 10_1	87.62	10.5														3	3	6	9	11	11	17	22	11	3	2	2
Number	Mixture	$\overline{\overline{X}}_1$ (in μ)	C_m (in percent)																									
1	$a + b$	59.26	8.7						3	9	38	38	21	41	36	5	5	2	2									
2	$a + c$	59.21	9.3				4	1	9	18	33	18	16	37	34	15	9	2	4									
3	$a + b$	58.72	10.8				2	0	6	11	43	27	24	40	33	5	5	2	2									
4	$a + e$	53.48	19.0	2	6	9	15	19	19	19	22	13	11	23	28	5	5	2	2									
5	$a + f$	67.30	14.3						3	1	12	14	15	27	43	15	11	23	8	12	10	6						
6	$a + g$	69.89	14.4						1	1	10	13	12	23	30	7	13	20	28	18	13	3	4	3	1			
7	$a + h$	74.76	19.2						1	1	10	12	11	23	28	5	8	5	8	9	11	11	17	22	11	3	2	2
8	$e + g$	61.46	28.5	2	6	9	15	19	18	18	12	2	1	0	2	2	8	18	23	18	13	3	4	3	1			

Test VI (tables 18, 19).—Mixing spectra of *P. contorta* 5 with spectra of various other different species of pine, and also the widely different spectra of *P. aristata* 1 and *P. sabiniana* 4 (table 18, bottom row) shows a consistently larger C_m than that of *P. contorta* 5, except in one case (table 19, mixture 1). The means of individual components in the mixed samples are all significantly different from

TABLE 19

STATISTICS OF VARIOUS MIXED SAMPLES OF TWO COMPONENTS AS IN TEST VI
(N = 200)

Mixture number	Species and sample number	$\overline{X}_m$ (in μ)	SD_m (in μ)	Class number	C_m (in percent)	Range	Remarks
1	*P. contorta* 5 + *P. murrayana* 15	59.26	5.17	11	8.73	47–77	Class-C indexes are normal, and the mixed nature of the sample cannot be detected by means of the indexes.
2	*P. contorta* 5 + *P. muricata* 7	59.21	5.52	12	9.32	41–77	
3	*P. contorta* 5 + *P. balfouriana* 8	58.72	6.32	13	10.76	51–77	
4	*P. contorta* 5 + *P. aristata* 1	53.48	10.16	16	19.00	32–77	Mixtures have abnormal Class-C indexes, and leave no doubt about being mixtures.
5	*P. contorta* 5 + *P. lambertiana* 7	67.30	9.64	14	14.32	47–86	
6	*P. contorta* 5 + *P. sabiniana* 4	69.89	10.04	17	14.36	47–95	
7	*P. contorta* 5 + *P. jeffreyi* 10	74.76	14.38	20	19.24	47–104	
8	*P. aristata* 1 + *P. sabiniana* 4	61.46	17.50	22	28.47	32–86	

NOTE: Mixtures 1–7 all involved *P. contorta* 5. The mean of *P. contorta* 5 is the larger of the two in mixtures 1–4, but is the smaller in mixtures 5–7. Mixture 8, considering *P. aristata* 1 in the position of *P. contorta* 5, is also the smaller of the two.

one another. This confirms the conclusion from Test I, result no. 1. The exception: mixture of *contorta* and *murrayana* is covered by statement Test I, result no. 4.

Until now the problem has been to determine why there is an increase in C_m in a mixed sample. The conclusion is that an increase in C indicates a mixture of species. But how much of an increase can be attributed to mixing? Since in fossil studies the original C values are never known, one cannot determine that a certain value is an increased value unless one knows that it is not a normal value. Note that in the reference samples (appendix table 1) C varies from 7.3 to 13.8 per cent. From the tests, we also know that C increases with increasing number of classes. There is a relationship between number of classes and C. For a total of

246 samples, the Pearson's coefficient of correlation between class number and *C* is 0.6305, which is significant at the 0.001 level.

It is therefore valid that a composite index for class number and *C* be used as a measure of homogeneity. The composite index is expressed by combining the two values into a hyphenated number. The index 10.8–9.7 describes a spectrum having 10.82 classes and a *C* of 9.68 per cent. The average class-*C* index for 246 reference samples is 10.5325–9.8211, abbreviated to 10.5–9.8 and in the tables further simplified to 10–10.

Among the individual samples of the different species in table 11, the class and *C* values are both low in *P. remorata, P. contorta,* and *P. murrayana,* both very high in *P. jeffreyi* and *P. coulteri.* A combination of low class number with high *C* is found among the *P. albicaulis, P. flexilis, P. aristata, P. balfouriana,* and *P. monophylla* samples. These are all pines having pollen grains with sculptured distal membranes. *P. lambertiana* has a medium class number and a high *C*. All the rest have a class number and *C* that are both medium.

In table 11, column $SD_T/\overline{\overline{X}}_T$ gives coefficients of variance of the species as a whole. The values of SD_T are those found from the total sum of frequencies of samples within one taxon. Thus *P. albicaulis* has nine individual samples. By adding these spectra into one, we produce one "mixed" spectrum of one species. Since these spectra are from one species only, it is reasonable to assume that they are from one universe, or, in other words, that the "mixed" spectrum we produced is actually a pure one. If the assumption is true, the coefficient of variance found should be small, which is precisely what we have found. The *C* thus calculated for *P. albicaulis* is only 5.3 per cent, whereas the average of the individual *C's* of 9 samples of the same species is 11.5 per cent, very much higher.

The procedure is to compare a class-C index of an unknown sample with those of the references (appendix tables 1, 2). A normal index shows high homogeneity of the sample and will be found to correspond to one or a number of the references. An abnormal one will not be comparable to any of the references.

While heterogeneity indicates mixing of universes, homogeneity does not always indicate pure samples of one species. Homogenity may be brought about by samples of similar statistical makeup, such as samples with mean not significantly different or with similar class number and range. It may also be brought about by averaging certain samples with large differences in *C*, or in their sample size.

I have dealt so far only with mixture of two components. Identification of these and mixtures of more than two components are discussed later (see section on class-C index).

Part III

SPECIFIC IDENTIFICATION OF SINGLE SPECIES

Introduction

The process of identification is essentially one of elimination. The needed data that have been gathered and calculated are arranged in various forms convenient for use, such as tables, diagrams, keys, and so on, preferably in the order of the process of elimination. I have assembled many different tables for my own use, but I have included in this study only two master tables: appendix tables 1 and 2. Appendix table 1 contains the data on all reference samples of mean values of corpus, saccus, and overall sizes, the ratio, the cos A index, together with the composite class-C index and the standard deviation of the mean of the corpus. Appendix table 2 contains individual spectra of corpus measurements on all reference samples. I have also included two much simplified tables (tables 10, 11) on modal pattern, range, and class-C index, simply for convenience of reference while discussing these statistics.

The following examples show how the identification of an unknown sample is worked out. The first one is given in detail to acquaint the reader with the process. The others are abbreviated.

Example I

The process of identification of an unknown from references is mainly a process of comparison of data and elimination, and can be accomplished in many ways. In the first example the process follows a prescribed order, which I found to be convenient and which has yielded the best results.

The statistics are considered in the following sequence:

1. The C is first examined to determine homogeneity of the unknown.
2. The class-C index and the range are compared with the references from which a list of eligibles is obtained.
3. The modal distribution is studied and compared with those of the references.
4. Elimination by mean values of the known within a range of mean plus and minus four times standard error.
5. Elimination by ratio indexes.
6. Elimination by cos A index.

Morphological features are usually considered at the very beginning. In this example I deliberately left the morphology out of consideration to give the statistics a more rigid test. The data are as follows (table 20):

STATISTICS

Range: 38μ–74μ.
Class number: 13.
Mean or corpus breadth: 58.40μ; of saccus breadth: 40.52μ; of overall breadth: 73.82μ.
Spectrum: 1, 1, 0, 1, 0, 17, 22, 20, 27, 9, 1, 0, 1.
Modal distribution: Major, class 62; minor, class 56, with co-mode, class 59.

Ratio indexes: Only one here, which is 1412.
Cos A index: 6153.
Standard error of mean of corpus breadth: 0.51μ.
Coefficient of variation (*C*):8.7.
Class-C index: 13-9.

MORPHOLOGY

Distal membrane psilate.
Pattern of cappa: *Y*.
One-third of the grains have thick crests.

IDENTIFICATION

(Table 20)

The *C* of this sample is 8.7, which is smaller than that of Group A pines but normal for species of Group B pines (table 11). Since the value is normal for a single species, the sample is taken for granted as homogeneous and is believed to be composed of pollen grains of one species.

We obtained from the references (appendix table 1) a list of possible species of similar range and similar class-C index. For the range of 38μ–74μ list all samples and species under a range 38μ ± one class and 74μ ± one class. For class-C index, list all samples and species that come under the indexes: 12-9, 13-9, 13-8, 13-10, 14-9. Compare the modal distribution of the unknown with known species, using the diagrams (fig. 13). We find that a bimodal distribution with a 62 major and a 56 minor mode and co-mode 59 occurs in samples of *P. monticola, P. contorta, P. ponderosa, P. attenuata,* and *P. murrayana.* I found this particular modal distribution in only two of my samples: *P. murrayana* 9 and *P. attenuata* 14. Though the bimodal combination also occurs in samples of *P. flexilis* 2, *P. monophylla* 11, and *P. attenuata* 17 and 19, most of these are eliminated by the range of mean. Under ratio I found only two samples having a ratio of 1412. These are *P. contorta* 14 and *P. murrayana* 27A.

In ordinary processing we need only investigate the cos A of these last two samples to conclude that the sample is *P. murrayana.* For this example, we may compare all the cos A values of samples considered as eligibles after elimination by mean (see table 20, eligible samples in italics in column 5, their cos A indexes in the last column). We find that the cos A of *P. murrayana* 27A (6301) is closest to the cos A of the unknown (6153) and that the cos A of *P. murrayana* 9 (6573) is the next in line. *P. contorta* 14 was the only other sample left with *P. murrayana* 27A under the column ratio, but its cos A (7250) is very distant from the unknown (6153).

COMMENTS

1. The identification can be made on the basis of the ratio index alone, for a ratio of 1412 is extremely rare among the samples. The unknown is actually a duplicate of the particular reference sample *P. murrayana* 27A. It is therefore listed now in the collection as *P. murrayana* 27B. There is a significant difference between the mean values of the samples 27A and 27B, but the ratio and the cos A index show that they are identical.

TABLE 20

IDENTIFICATION OF AN UNKNOWN BY ELIMINATION AS IN EXAMPLE I

Eligible species and sample number	Range	Class-C index	Modal distribution	Range of mean (in μ)	Modal in total distribution	Ratio index	Cos A[a] index
Unknown	38–74	13–9	62, 56–59	56.36–60.44	Bimodal (fig. 13)	1412	6153
Group A							
P. albicaulis	1, 11			11			7577
P. flexilis	6, 7		2	7			7063
P. monticola	2, 5, 9, 11			5, 9, 11	X		7436, 7201, 7398
P. lambertiana	1			1			7411
P. aristata	6A						
P. monophylla		1N, 14T	11	11			6913
P. torreyana	2			2			6779
Group B							
P. contorta	7A, 14	7A		7A, 14	X	14	6866, 7250
P. murrayana	5, 9, 16, 21, 27A 37		9	9, 16	X	27A	6573, 6975, 6301
P. ponderosa	9, 14, 34N	4, 28, 29, 34		9, 34N	X		6892, 7199
P. attenuata	3, 10, 14, 16, 20 21, 23, 24, 29A	22	14, 17, 19	3, 10, 14, 16, 20, 21, 23, 24	X		7170, 7537, 7476, 7321, 7490, 6732, 6968, 7146
P. muricata	1B, 2, 3, 7	2, 5, 11, 13		1B, 2, 3, 7			6763, 6740, 6762, 7184
P. radiata	2, 3B	5		2, 3B			7864, 7854
P. coulteri		3T					
P. jeffreyi		8, 10, 12, 15, 16, 21, 26, 26					
P. remorata			4				

[a] All Cos A indexes entered below the unknown are in order of those of species and samples remaining eligible after elimination by range of mean as listed under range of mean and Ratio index.

2. If the morphological features are taken into consideration to begin with, the scope of search is much narrower, for then the pines of Group A would be eliminated at the start. *P. muricata* and *P. radiata* would be eliminated by the pattern of the cappa. We would have to consider only *P. contorta, P. murrayana, P. ponderosa,* and *P. attenuata.* Among these *P. murrayana* is the one that could possibly have many pollen grains with thick crests. We arrive at the same conclusion as before. Since "thick crest" is not considered a diagnostic feature, it is safer to use the statistics than morphology, though in this case the same correct answer may be reached.

3. The mode is another way of describing the central tendency. It may not necessarily be used if the mean value is sufficient for the identification. In this case we have a "disconformity" between the findings from range and class-C index, and it appears advisable to consult the modal distribution as well.

EXAMPLE II

MORPHOLOGY

The pollen grains all have psilate distal membrane. The pattern of cappa is Y.

STATISTICS

Range: 44μ–80μ.
Mean of corpus breadth: $65.48\mu \pm 1.18\mu$; of saccus breadth: 53.72μ; of overall breadth: 92.90μ.
Modal distribution: unimodal at class 65.
Ratio indexes: 1214, 1213, 1113, 1114.
Cos A index: 7242.
Class-C index: 13-9.

IDENTIFICATION

(Table 21)

The listings of range, class-C index, and modal distribution after elimination by range of mean, 63.12μ–67.84μ, include 6 species and 17 samples as eligible (see table 21 under ratio). The cos A index of the unknown (7242) is closer to 3 of the *P. ponderosa* samples (7314, 7343, 7368) than to any other species. The unknown is identified as *P. ponderosa washoensis.* The method detects the species correctly but not the variety, which statistically is identical with *P. ponderosa* (see also Appendixes III, IV).

EXAMPLE III

This example serves to show that when a certain statistic appears to be very characteristic, we may skip a number of steps.

Pollen grains have psilate distal membranes. The pattern of the cappa is Y. The sacci are unusually large, and in many instances larger than the height of the corpus.

Among the statistics (table 22), the mean value of corpus, $76.49\mu \pm 1.17\mu$, is very characteristic indeed, for not many pines have such dimensions. I therefore go directly into a table listing mean values. (This table is not included here, but all

TABLE 21

IDENTIFICATION OF AN UNKNOWN BY ELIMINATION AFTER ELIMINATION BY RANGE OF MEAN 63.12μ–67.87μ AS IN EXAMPLE II

Eligible species and sample number	Sample numbers of eligible species after elimination by range of mean, 63.12–67.87 μ				
	Range	Class-C index	Modal distribution	Ratio	Cos A index[a]
Unknown......	44–80	13–9	Unimodal: 65	1214, 1213, 1114, 1113	7242
P. ponderosa...	4, 14, 33, 28 29A	4, 28	16, 17, 18, 28	4, 14, 16, 28, 29A, 33	7343, incomplete, 7009, 7368, 7062, 7314
P. attenuata....	13, 22, 29A	22	18, 22, 29A, 33	13, 22, 29A, 33	6802, 7063, 7501, 7114
P. radiata......	5, 6, 1	5	5, 6	5, 6, 1	7025, 7027, 6403
P. sabiniana....	......	....	2A, 3	2A, 3	7036, 7694
P. monticola ...	7	....	......	....	(6664)[b]
P. torreyana....	......	....	5	6	(6650)[b], 6786
P. muricata....	......	....	14	....	(7184)[b]
P. murrayana ..	......	....	31	31	6830

[a] The cos A indexes of species below the unknown are in the order of samples listed under Ratio.
[b] Indexes in parentheses were eliminated by Ratio, but given for comparison.

TABLE 22

IDENTIFICATION OF AN UNKNOWN BY MEANS OF C AND COS A OF ELIGIBLE SPECIES AS IN EXAMPLE III

(Means in μ)

Species and sample number	C (in percent)	Cos A	$\overline{X}_1$	$\overline{X}_2$	$\overline{X}_3$
Unknown...............	7.63	6691	76.49	102.56	62.24
P. jeffreyi X *ponderosa* 22	7.63	6712	76.25	102.56	63.02
P. sabiniana 4...........	6.40	6789	77.87	105.50	63.68
P. coulteri 6.............	7.64	6887	75.68	103.70	62.78
P. coulteri 10............	8.90	6604	76.36	101.45	61.80
P. coulteri 11............	9.97	6528	76.89	101.42	63.20

mean values are in appendix tables 1, 2.) Among the species having a mean value within the range of mean between 74.15μ and 78.83μ are:

P. coulteri 1, 5, 6, 10, 11
P. jeffreyi 2, 9, 12
P. jeffreyi × *ponderosa* 22
P. sabiniana 4

The ratio index of this sample is 1213 (mean of saccus = 62.24μ, mean of overall = 102.56μ). Under this we eliminate all of *P. jeffreyi* and *P. coulteri* 1 and 5.

A comparison of the cos A of the remaining species and samples (table 22) indicates that this unknown is *P. jeffreyi* × *ponderosa,* which is the correct answer. It is obvious that cos A acts as a discriminating function among otherwise inseparable universes.

Notice that we skipped a number of steps and went right on to the ratio because the mean value happened to be an extreme one.

Example IV

MORPHOLOGY

The unknown has verrucose distal membrane, with many abortive grains.

STATISTICS

Three lots, *A, B,* and *H,* were drawn. Each lot consists of an initial measurement of 100 grains. A succession of mean values of various lot sizes above 100 were also taken. Lot *A* at 100 grains has a mean corpus breadth of 50.96μ. At 160 grains the mean size is 50.94μ. The modes remain the same throughout, being 53, 44, 62 in order of dominance. Only *P. balfouriana* among all samples has this trimodal distribution.

Lot *B,* measured some days later, gives a larger mean of 54.38μ. The modes are the same as those of lot *A.*

All the above lots do not include the abortive grains, if included, they depress the mean value to 48.74μ and the modes become 50, 41, 62—obviously a shift toward the smaller end. No such modal combination has been recorded among the references.

Although the mean of lot *B* is significantly different from that of lot *A,* the ratio indexes are the same: 1112, 1113, 1213, 1212; the *C* is also similar: 12.25 per cent and 12.00 per cent; and the cos A is 6218 compared with 6481 of lot *A.*

Amidst the many eligibles found under the ratio indexes, the cos A of the unknown is closest to those of *P. balfouriana,* whether that of lot *A* or lot *B* is used for comparison (table 23).

Pollen lot *H,* collected from the same herbarium sheet, gives a mean of 51.32μ, ratio indexes of 1012, 1112, 1013, 1113, a cos A of 6640; the *C* is 10.3 per cent. The data also indicate *P. balfouriana.*

The identification is correct. In fact, the unknown sample lots *A* and *B* and *H* came from the same locality as *P. balfouriana* 6. If we had used the data that included the abortive grains, we would not have been able to identify the unknown as any one of the existing pines in California. This shows that immature grains should be excluded from consideration.

The fact that these three lots were derived from the same herbarium sheet and yet give three mean values that are significantly different from one another shows that the variation of pollen size is beyond the confidence limits of the mean of one measurement. The fact that by considering three measurements simultaneously, rendering them into ratios and cos A index, we can arrive at the true identity of the unknown in spite of the difference of the mean, strongly indicates the validity of these measures.

Example V

Five samples of pine pollen grains were supplied by Professor Ledyard Stebbins as unknowns for determination in April, 1963. All five were correctly identified. The pertinent data for these five samples are given in table 24.

Sample 1. *Pinus murrayana*

Identification is immediately determined from the class-C index of 7-7.0, and a psilate distal membrane with a mean of the corpus 58.70μ ± 0.82μ. The only eligibles listed under these features are *P. murrayana* 15, 22, and *P. murrayana*

TABLE 23

Identification of an Unknown by Cos A of Eligible Species as Used in Example IV

Eligible species and sample number	Cos A	Eligible species and sample number	Cos A
Unknown lot *A*	6218	Unknown lot *H*	6640
Unknown lot *B*	6481		
P. monticola 4	6776	*P. monticola* 4	6776
P. monophylla 5	6658	*P. aristata* 7	7032
P. monophylla 12	7060	*P. aristata* 8	7114
P. balfouriana 2	6596	*P. balfouriana* 2	6596
P. balfouriana 4A	6480		
P. balfouriana 4H	6254		
P. balfouriana 6	6209	*P. balfouriana* 6	6209
P. albicaulis B7	7564		
P. flexilis 2N	7258		
P. flexilis 7	7063		

var. *latifolia* 9. From the ratio index, four more numbers of *P. murrayana* are found as eligibles. From the range of the sample, *P. murrayana* 17 and *P. contorta* 18 are both indicated. The weight of evidence readily shows that the sample is *P. murrayana*. The sample *P. contorta* 18 is from San Juan Islands, Washington, and in its statistical aspects is not separable from *P. murrayana* of California.

Sample 2. *Pinus flexilis*

The identity is found by the cos A. Elimination through the class-C index, the range, the mean, and the mode and the ratio index gives us three eligibles—*P. aristata* 6 (cos A = 7588), *P. flexilis* 2N (cos A = 7470), and *P. balfouriana* 18 (cos A = 6713). The cos A of the unknown is 7403. It is obviously not *P. balfouriana*. The cos A of the unknown differs from that of *P. aristata* 6 by 185, whereas it differs from *P. flexilis* 2N by only 67. This difference, coupled with the consistency of cos A of *P. flexilis* known from references, indicates that the unknown is *P. flexilis*.

Sample 3. *Pinus jeffreyi*

Pollen of this species is easily recognized by the pattern, by its thickness of crest, and by its large size. Statistical data do not give the identification in this case because the cos A of the sample resembles that of *P. coulteri* 5 more than that of

TABLE 24

IDENTIFICATION OF FIVE UNKNOWN SPECIES OF *Pinus* AS IN EXAMPLE V

Unknown sample	Species identified	$\overline{X}_1$	$\overline{X}_2$	$\overline{X}_3$	Class-C index	Range	Mode	Ratio indexes	Cos A	Group according to distal membrane
		(in μ)								
1	*P. murrayana*	58.70	79.67	46.94	7–7.0	50–65	56, 62	1213	6824	B
2	*P. flexilis*	53.09	78.20	50.48	11–12.5	36–68	50, 56, 62	1014, 1015	7403	A
3	*P. jeffreyi*	74.14	109.87	68.30	16–12.0	50–98	74, 65, 89, 56	1014, 1015, 1114, 1115	7489	B
4	*P. coulteri*	73.52	103.04	59.90	9–6.2	62–116	74, 77	1114, 1214, 1113, 1213	7128	B
5A	*P. monophylla*	60.74	81.86	47.48	12–8.4	44–71	65, 57, 62	1213, 1313, 1312	6861	A
5B	*P. monophylla*	61.19	82.94	47.12						A
5C	*P. monophylla*	60.00	.....	47.00						A

P. jeffreyi. Since it was evident that the sample was *P. jeffreyi* because of its crests, the real identity of reference samples *P. coulteri* 5 was in doubt. The actual reason for the incompetence of statistics in this case probably lies in the need for a larger sample size than the standard 100 used for pollen size above 70μ.

SAMPLE 4. *Pinus coulteri*

The sample is recognized by the large size of the corpus, the persistent globular shape of corpus and saccus, and the fine texture of pattern, though it has the same *Y* pattern as *P. jeffreyi.* No indication of identity is obtained from the class-C index, the range, or the mean size of corpus. The mode alone indicates *P. coulteri* exclusively. The cos A is different by 96 from *P. coulteri* 5.

Since this *P. coulteri* is identified by a comparison of the unknown with reference sample 5, suspicion of its correct identity raised in the study of sample 3, above, is not justified.

Reflection on the study of these two samples, nos. 3 and 4, indicates that the more powerful method of discrimination between these two species lies in the morphology rather than in the statistics.

SAMPLE 5. *Pinus monophylla*

Before this sample was studied, all reference data of *P. monophylla* did not give a mean of corpus anywhere near 60μ. The unknown sample has a large pollen size, above 60μ.

P. monophylla has a verrucose distal membrane that readily isolates it from many other pines. The white pines possess a large saccus frequently described as *Haploxylon* type, and a ratio between corpus and saccus of slightly more than 1.0. But *P. monophylla* has a small *Silvestris*-type saccus that yields a ratio of 1.1 to 1.2 or more. Since ratios in certain individual samples may overlap one another, it is not so conclusive a guide as the cos A. In the present example (table 24, sample 5A), the cos A is 0.68, which is typical of the *P. monophylla* group.

Based on the existing references, the statistics invariably lead to *P. monticola* because of the large mean size of the corpus of the unknown, rendering a final test by cos A not applicable. Either the size of the saccus or that of the corpus was then thought to be in error. Three separate measurements (table 24, samples 5A, 5B, 5C) show high consistency for these sizes, especially that of the saccus. This fact of small saccus, coupled with the cos A, shows that the unknown could not possibly be *P. monticola,* and that it must be *P. monophylla.*

From these evaluations it is obvious that statistical data do have their limits and should be supported by morphological characters, whenever available. The study shows that the cos A as devised is a more powerful means of identification than any other, though it does not work all the time. It also shows that for certain species, more reference samples are needed in order to give adequate coverage to the variation of the living taxa.

Part IV

SPECIFIC IDENTIFICATION OF MIXED SAMPLES

Introduction

The method of specific identification described above deals with spectra of a single species only. It is of value in studies of the modern species, and it would also be useful in the study of fossil pollen flora if it could be applied in the same way. The composite spectrum derived from measurements of fossil pollen in a sample consists of more than one species. It represents an average assemblage of pollen of many species from different trees of the same species, from different parts of the same tree, spanning many years of accumulation. The composite spectrum is therefore heterogeneous. The basic statistics, such as sample mean, coefficient of variance, cos A, modal distribution, and ratio index, which we have used so far, no longer represent one species but the aggregate of a number of species in different quantities. The statistics thus far are therefore not directly applicable to mixed samples.

It is my belief that the reference sampling will cover most of the statistical makeup of pollen grains of different samples and species. The statistical makeup of a mixed assemblage can be duplicated by pooling statistics on certain selected references. The basic problem is how to select these references.

CLASS-C INDEX

I have previously discussed the use of the composite index of class number and coefficient of variance (C) as an indicator of homogeneity of a sample (see above, Coefficient of Variance). In the case of the 20 pines given in Part I (see Statistics) the class-C index is 23-17.5. None of the reference samples has this index. Among the references, the maximum class number is 20 and the maximum C is 16.6 per cent found in *P. jeffreyi* (table 9).

In the identification process the eligibles would be eliminated by the range of the mean value. Hence, even if there is a sample with a large a class-C index as 23-17.5, the sample must have a compatible mean value in order to remain eligible. Since the 23-17.5 index found no match in references, the sample is believed to be a mixture and heterogeneous.

There is, however, no way to tell from the class-C index how mixed the sample is, for even two spectra of extreme differences in mean and in statistical makeup could give a very high class-C index. For instance, in Test IV, mixtures 4, 5, and 6 or two components resulted in class-C indexes of 20-18, 22-22, and 24-27, respectively (table 16). Again in Test VI, mixture of *P. aristata* and *P. sabiniana* gives a class-C index of 22-17.5 (table 19). For mixtures of two components only, these are very high indexes.

There appears to be no correspondence between *C* and the number of components in a mixture, for the index reflects simultaneously two conditions—the number of components and the span of the spectrum—and it would not be easy to say offhand which condition decides the value of a certain index. Theoretically speaking, for any given class number (span of spectrum), there is a normal range of *C* with respect to the mean value (of corpus breadth). A sample with a certain mean and a certain class number, but with a *C* outside that normal range, is therefore abnormal.

In identification of an unknown, with given statistics and spectrum, one examines the class-C index and tries to match it with existing references. If no match can be found, the index is then regarded as abnormal, and the unknown is interpreted as mixed. Even if a match from reference is found, the unknown sample can still be a mixed one because of the following conditions:

1. The means and standard deviations of mean of the components are very similar, therefore, giving similar normal *C* (table 16, Test IV, mixture 1).
2. The components have significantly different means and different variance in such a way that class-C index of their mixture remains normal. This case is hypothetical, and is very rare in reality.
3. The range of one of the components falls within that of the other components, and mixing does not increase the class number (table 16, Test IV, mixture 1; table 19, Test VI, mixture 1).
4. The spectrum of the unknown may be longer than the originals because of mixing, but not more than two classes from the original longest of the components (table 19, Test VI, mixtures 2, 3).

The apparent homogeneity of a mixed sample of two components does not always prevent the sample from being recognized as heterogeneous if other statistical data and morphology are also considered. These cases are rare. Take, for example, the dense cluster of samples from existing reference samples between the ranges of 44 to 50μ and 74 to 80μ, and class-C indexes 10-10, 11-9, 11-10, 11-11, and 12-10. Listed under these limits of range and class-C indexes are 98 samples, or 40.0 per cent, of the total 246 samples of *Pinus*. Sixty-one samples are listed under the class-C index and 60 samples are listed under the range. Of these 121 samples, only 23 occur on both lists (table 25). If any 2 of these 23 samples were mixed, they would probably produce a sample with an apparently homogeneous class-C index. These 23 samples, however, represent only 9 species: 3 belonging to the Group A pines and 6 to the Group B. Among the Group A's, 1 has a Y pattern and 2 have the W pattern. Among the Group B's, 2 have Y pattern, 2 have W pattern, and 2 have WY pattern. The total number of combinations, excluding the possible mixtures within species, is 224. Mixtures of A and B groups, 90 cases, should offer no problem of being recognized as mixtures, and these should also be deducted from the total, thereby leaving 134 cases to be evaluated. Of these, only 27 cases pose the difficulty that two samples of the same group and the same pattern may be mixed. Hence, even at the densest cluster of samples, the homogeneity of a mixture of two yields very few possible cases that may hinder recognition by means of the class-C index. The chances of getting a homogeneous, but mixed, sample in ranges and class-C indexes other than those above are even more rare. The chances of getting an apparently homogeneous mixture of 3 components are practically nil.

Test IV gives the mixtures of two spectra, *P. attenuata* and *P. torreyana*, in various ways (table 16). In mixture 1, the class-C index and the range are no different from the original, for there are no significant differences between their means. Their standard deviation of mean is practically identical. The mixture, to all appearances, especially the unimodal distribution, suggests that of a homogeneous sample of a single species. Identification will of course proceed as if the sample is indeed of a single species. Under range of 59μ to 89μ one finds 13 individual reference samples, of which, after elimination by the mean of corpus breadth, only three eligibles remain: *P. lambertiana* 6, *P. ponderosa* 18, and *P. torreyana* 6T. Under the class-C index, one finds five samples, of which, after elimination by mean, only one is left: *P. torreyana* 6T. As far as the identification goes, this will be the answer, and we will have missed identifying *P. attenuata.*

TABLE 25

OCCURRENCE OF SAMPLES SIMILAR IN RANGE AND CLASS-C INDEX WITHIN THE RANGES 44μ TO 50μ AND 74μ TO 80μ, AND WITH CLASS-C INDEXES 10–10, 11–9, 11–10, 11–11, AND 12–10

Species	Numbers of samples	Group according to distal membrane	Cappa pattern	$\overline{X}_1$ (in μ)	$\overline{X}_2$ (in μ)	$\overline{X}_3$ (in μ)	Ratio index	Cos A
P. flexilis	1		W	55.2	51.8	79.4	1014	0.72
P. lambertiana	2	A	Y	64.6	65.8	95.4	0914	0.74
P. monticola	2		W	59.0	54.8	84.2	1015	0.71
P. ponderosa	4		Y	62.8	56.8	90.7	1114	0.73
P. attenuata	6		YW	62.1	54.1	88.9	1114	0.72
P. murrayana	1	B	Y	57.3	46.4	78.1	1213	0.68
P. radiata	3		W	62.7	55.3	90.2	1124	0.73
P. muricata	1		W	62.0	54.4	87.2	1114	0.71
P. sabiniana	3		YW	68.4	58.6	97.2	1114	0.72

We made the test difficult by not considering morphological features and other statistics that accompany the problem. Yet we managed to obtain a partially correct answer. The result is not at all surprising because an apparently "homogeneous" but really mixed sample may bear some resemblances to one of the components.

If one takes into consideration the omitted features, the mixed nature of the spectrum will be evident. First of all, the existence of two morphologically different pollen types in the sample will amply demonstrate that it is a mixture. Second, the *cos A* of the unknown does not come anywhere near the normal limits of that of *P. torreyana* 6T. The ratio and the modal distribution are also different.

To find the other component in the example above is no problem, but it is much too ideal a case and seldom, if ever, occurs in nature. Pollen grains from each component in real samples are apt to be unequal in number. With a normal class-C index, the best attempt will probably always result in identifying only one of the components.

Mixtures 2 to 6 of the same two spectra (table 17) were achieved by shifting one of the spectra to a higher and the other to a lower class, one class at a time.

TABLE 26

COMPARISONS OF MIXED SAMPLES OF THREE COMPONENTS AS IN TEST VII

Mixture	Components A + B + C	N	$\overline{X}_m$ (in μ)	SD_m (in μ)	SE	Class number	C (in percent)	Conditions
1	*P. contorta* + *P. murrayana* + *P. muricata*..........	300	60.24	5.31	0.61	13	8.82	$\overline{X}_A > \overline{X}_m > \overline{X}_B > \overline{X}_C$
2	*P. contorta* + *P. murrayana* + *P. lambertiana*........	300	64.51	8.55	0.99	14	13.25	$\overline{X}_C > \overline{X}_m > \overline{X}_B > \overline{X}_A$
3	*P. aristata* + *P. muricata* + *P. sabiniana*..........	300	60.71	15.45	1.76	23	24.24	$\overline{X}_C > \overline{X}_m > \overline{X}_A > \overline{X}_B$

NOTE: For spectra, see table 18; $\overline{X}_m$ is mean of mixed sample.

These "original" components are, in fact, no longer *P. torreyana* or *P. attenuata,* but new spectra created out of them. One would not expect to find their identities by single-species identification. Trial will show, however, that in mixture 2, in spite of the changes—an extension of two classes in spectrum, and a larger *C*—the names of *P. torreyana* and *P. attenuata* still come up among the final four eligibles.

No eligibles can be found for mixtures 3 to 6. These are obviously heterogeneous from their class-C indexes. There exists no such pines.

Test VI (table 19) shows the mixing of the spectrum of *P. contorta* 5 with various other spectra of different species. The differences between the mean values of all these species and that of *P. contorta* 5 are significant. The original *C*'s are also varied. It would not be difficult to recognize mixtures 4 to 8 as heterogeneous because of their abnormal class-C indexes, but it would be impossible to recognize mixtures 1 to 3 as heterogeneous solely by their class-C indexes since these are apparently normal. In fact, for mixture 1, *P. contorta* and *P. murrayana* are not always distinguishable from each other when single species, much less when they are mixed. For mixture 2, *P. contorta* differs from *P. balfouriana* very clearly in morphology. There will be two types of pollen grains in the sample. For mixture 3, *P. contorta* differs from *P. muricata* in the pattern of cappa.

The chances of having a normal class-C index in cases of mixing three or more components are very small. The three examples in Test VII (table 26) show that in such cases there is no doubt of their heterogeneity. In mixture 1, the components are those indistinguishable in a mixture of two, as in Test VI (table 19). After mixing three together, the class number increases by two classes. The *C* is very low, being only 8.80 per cent, which would be considered normal for single species. But the references do not have a match for a combination such as class-C index 13-9 and compatible mean value of $60.24\mu \pm 0.61\mu$.

Mixture 2 (table 26) showed a mixing with a slightly wider spread. The classes have not increased from mixing, being 14, which is as long as one of the original components (see table 18, *P. lambertiana* 7). The *C,* however, is 13.2 per cent. From the reference data we again find no compatible fit for the description. Mixture 3 is made up of three widely scattered spectra so that the class number is greatly increased from 14 to 23 and the *C* is 24.2 per cent. This undoubtedly is a mixed assemblage. Hence it is clear that in any mixing of more than three components the class-C index can be depended on as a discriminant.

ESTIMATION OF THE NUMBER OF COMPONENTS

In sampling references, a sample size (*N*) of 100 individuals is used as a standard though other *N* sizes are also used (e.g., 50). In mixed assemblages, a sample size of 100 is sufficient in most cases. Even this sample size may not easily be reached in fossil samples. The indexes that are most affected by sample size are the class number and range of spectrum, but in practice these effects are minimized in identification, by taking into consideration references within one class higher or lower than that of the unknown. The sample size also affects the standard deviation, but the effect is relatively small if the sample size is between 50 and 300. Only when the sample size is too small does it affect the mean value. There is serious concern when the effect on standard deviation and mean value accentuates or diminishes the *C* more than 1 per cent each way, but this rarely occurs.

An initial sample of 100 grains is sufficient basis for the purpose of estimating the number of components. Obtain the mean, the variance, and coefficient of variance first. When the *C* indicates mixing, inspect the spectra of the corpus, saccus, and overall measurements. Note the locations of the modes and quantities of each individual class of the spectra, and count the number of classes.

From the reference data (appendix tables 1, 2) the normal length of a spectrum (range and number of classes) is easily found. A mixed spectrum is usually longer, and multimodal, and has unevenly distributed frequencies among the classes. Experience from studying the references will help in spotting the obvious at the outset, and peculiar modal combinations may give a clue to the number of components.

The number of components can be estimated from morphological considerations. Sculpture of the distal membrane and sculpture of the cappa are two useful guides. Estimation can also be made and adjusted during the process of identification.

Mathematical Considerations

The following is an example designed to show the elements in mixed-assemblage differentiation. I have limited the components for the mixed sample to three for the sake of convenience. Our aim is to find the kind and the amount of pollen grains of each component in this mixed sample.

Study pollen samples of single species one by one, obtaining from them all the data needed. These data constitute the references. For every sample, group the measurements (corpus breadths, for instance), and enter them in the various classes to which they belong. This frequency distribution is a spectrum. Now take three spectra, each of a reference sample, and each representing one species, (*A, B, C*) and add them together class by class. We then have a mixed spectrum of three components. Label this spectrum I. Calculate the average of means, $\overline{\overline{X}}_{\text{I}}$.

Next, mount a slide on which one drop of liquid containing pollen grains of each species chosen (*A, B, C*) are mixed. Draw a sample of 100 grains from this slide. The measurements of these 100 grains will give us a second spectrum, designated as spectrum II and its mean: $\overline{X}_{\text{II}}$. Obviously spectrum I contains the same three components as spectrum II, but the quantities of each component may not be of the same proportion. For manipulation, multiply the frequencies of spectrum II by 3, to give it a sample size of 300. Spectrum I is made up of three components in equal shares (i.e., 100 grains each of species *A, B,* and *C*), whereas spectrum II may or may not have components of equal shares. The deviation of the mean $\overline{\overline{X}}_{\text{II}}$ from $\overline{\overline{X}}_{\text{I}}$ measures how unequally drawn from the mixture sample II is. The total sample size of both spectra is 300. If $\overline{\overline{X}}_{\text{II}}$ is the larger, then the individuals having a mean value below $\overline{\overline{X}}_{\text{I}}$ must be less than 100 per sample, and those with mean values higher than $\overline{\overline{X}}_{\text{I}}$ must contain more than 100 grains per sample. Should the value of $\overline{\overline{X}}_{\text{II}}$ be smaller than $\overline{\overline{X}}_{\text{I}}$, then the reverse is true. In the reference samples, the sample size (N) was equal, being 100 grains for each species, so a total of 300 grains makes up the composite spectrum. The sum of all three products of sample size and mean is therefore equal to 300 times $\overline{X}_{\text{I}}$.

Let x, y, and z be the sample size of each individual, and let A, B, and C be the means of component samples A, B, and C. The following equations may be formulated:

$$Ax + By + Cz = N\overline{\overline{X}}_{\text{I}}, \text{ where } x = y = z = 100; \quad (1)$$

$$Ax + By + Cz = N\overline{\overline{X}}_{\text{II}}, \text{ where } x, y, z \text{ are not necessarily equal in number, since } \overline{X}_{\text{II}} \text{ is not equal to } \overline{X}_{\text{I}}; \quad (2)$$

and

$$x + y + z = 300. \quad (3)$$

The number of solution to equations (2) and (3) is infinite if all real numbers are considered. In our problem there are some limitations. First of all, only positive numbers are considered. Second, x, y, or z can never be zero, otherwise we would not have three components. Third, the variations of one of these, say z, depends on the others, x, and y; and the limits are set by the values when $x \to 0$ and $y \to 0$. Fourth, though we cannot assume that all these numbers must be integers, nevertheless we do not work with fractions or square roots in pine pollen. The integral solutions provide skeleton models for our solutions.

There should be only one correct solution to the problem. To find it, we need some other relationships that can be put into a third equation, besides (2) and (3), yet so far I have not been able to formulate one. It is through iteration that the solution has been achieved, and such a solution must be regarded as an approximation of the true one. Further work is therefore needed in this direction to find the solution.

Some other relationships may be noted from the spectrum. The spectrum is a composite resulting from adding class by class the frequencies of each of the components. Each tail of the spectrum most probably represents the frequencies of only one of the components. The total sample size of that particular component bears a yet undetermined relation to the size of this class. The modes, especially those significantly large in number, must be interpreted as the sum of frequencies of three spectra at this particular class. It follows that in iteration, or spectrum-fitting, due consideration must be given to the distributional pattern of each of the three components from their references.

Added to the limitations stated, there is one requiring that the fitted individual spectra have mean values identical with that of the reference. The sum of frequencies of every class should be equal to that shown by the unknown spectrum.

These considerations narrow down the possibilities from a multitude of solutions to a closely approximated one.

AN ILLUSTRATION

Before working on a real problem of unknown pines, let us study a sample problem to illustrate the process of spectrum-fitting.

Let us assume that the three components A, B, and C have means 4, 3, and 2, respectively. Their average is therefore 3 ($\overline{X}_{\text{I}}$). The total sample size, being 100 each, is 300. If the mean ($\overline{X}_{\text{II}}$), of a composite spectrum II, drawn from a mixture of the three components, is 2.5 instead of 3, it is obvious that component A with mean of 4 must have been drawn less frequently than 100, and that component C with mean of 2 must have been drawn more often than 100. The intermediate component becomes a dependent variable. It may be drawn less or more depending on the numbers of the extremes.

Substituting into the equation (2), we get

$$4x + 3y + 2z = 300(2.5) = 750. \tag{4}$$

Dividing both sides by 2, we obtain

$$2x + 1.5y + z = 375.$$

Solving this equation simultaneously with (3), we have

$$x + 0.5y = 75. \tag{5}$$

Rendering this into parametric equations, then

$$x = 75 + t$$
$$y = 2t$$
$$z = 225 - t \text{ where } 0 < t < 75.$$

The limitation set in this case is that the solution lies in the range bounded by the relation $0 < t < 75$. Thus when t is zero, $x = 75$, $y = 0$; and when t is 75, $x = 0$, $y = 150$. The number of integral solutions then is only 74, when t is an integer, and positive. Other solutions involving noninteger values of t will be located within these limits.

We now proceed with spectrum-fitting.

On the work sheet (table 27) are given the spectra of three reference individuals (A, B, C) and their composite spectrum I. Drawing from a mixture of the three components of spectrum I, where N is the same for all three, we form unknown spectrum II. Spectrum II differs from spectrum I in terms of individual class frequencies and mean, and therefore its components could not have been drawn equally.

Our fitting begins with finding the spectra of components by assuming $t = 0$, $t = 5$, $t = 10$, and so on, to $t = 75$, and labeling each successive set of findings of x, y, and z as hypotheses (*HO* 1, *HO* 2, *HO* 3, $\cdots$, *HO* 7).

HYPOTHESES 1 AND 7

In hypothesis 1, the frequency of $A5$ (component A, class 5) is 5. The frequencies of $B5$ and $C5$ are both zero. These are known from the reference spectra of A, B, and C. Also assumed are $A3 = A5 = 5$, $B2 = B4 = 30$, and $C1 = C3 = 50$. Since $A3 + B3 + C3 = 85$, $A3 = 5$, and $C3 = 50$, $B3$ must be 30.

Since $B3$ cannot be any less than 30, being the mode of the B spectrum, $B2$ and $B4$ cannot be more than 30. This implies that, knowing $A2 + B2 + C2 = 110$ and $A2 = 0$, $C2$ must be 80 at its *minimum*. Also, knowing $A4 + B4 + C4 = 50$ and $C4 = 0$, $A4$ must be 20 at its *minimum*.

Hence, when $B2$ and $B4$ are both 30, the quantities of the three components should be $x = 30$, $y = 90$, and $z = 180$ (x, y, z, are N's of A, B, C, respectively).

Since $B2$ and $B4$ can be as small as zero, if so, it follows that $A4$ must be 50 and $C2$ must be 110. This will be hypothesis 7 (*HO* 7) which constitutes, together with *HO* 1, the two extreme limits of our solutions.

TABLE 27

HYPOTHESES FOR SPECTRUM-FITTING

Sample	Component	Frequency by class midpoint: 1	2	3	4	5	N	$\overline{X}$ (or $\overline{X}$ for total)
Reference........	*A*	..	..	20	60	20	100	4
	B	..	25	50	25	..	100	3
	C	30	40	30	..	..	100	2
Total: spectrum I.....		30	65	100	85	20	300	3
Unknown: spectrum II..		50	110	85	50	5	300	2.5
HO 1............	*A*	..	..	5	20	5	30	4
	B	..	30	30	30	..	90	3
(Limit)	*C*	50	80	50	..	..	180	2
Total..........		50	110	85	50	5	300	2.5
HO 2............	*A*	..	..	5	25	5	35	4
	B	..	25	30	25	..	80	3
	C	50	85	50	..	..	185	2
Total..........		50	110	85	50	5	300	2.5
HO 3............	*A*	..	..	5	30	5	40	4
	B	..	20	30	20	..	70	3
	C	50	90	50	..	..	190	2
Total..........		50	110	85	50	5	300	2.5
HO 4............	*A*	..	..	5	35	5	45	4
	B	..	15	30	15	..	60	3
	C	50	95	50	..	..	195	2
Total..........		50	110	85	50	5	300	2.5
HO 5............	*A*	..	..	5	40	5	50	4
	B	..	10	30	10	..	50	3
(Answer)	*C*	50	100	50	..	..	200	2
Total..........		50	110	85	50	5	300	2.5
HO 6............	*A*	..	..	5	45	5	55	4
	B	..	5	30	5	..	40	3
	C	50	105	50	..	..	205	2
Total..........		50	110	85	50	5	300	2.5
HO 7............	*A*	..	..	5	50	5	60	4
	B	..	0	30	0	..	30	3
(Limit)	*C*	50	110	50	0	..	210	2
Total..........		50	110	85	50	5	300	2.5

Hypotheses 2, 3, 4, 5, and 6 are calculated in the same manner and the results are tabulated in table 27.

Notice that we start the estimation from one end of the unknown spectrum II. The frequencies at one end of a spectrum usually belong to only one of the components, so the quantity of the unknown spectrum in this end class is the same as that of the unknown component. Assuming also a normal distribution with a variance equal to zero and an equal spread of classes above and below the mean (= mode in the ideal case), we can estimate the frequencies of every class of the spectrum of this hypothetical component. Two components can be estimated in this manner because the spectrum has two ends.

TABLE 28
LIMITS OF x, y, z VALUES OF COMPONENTS A, B, C, AND INTERPOLATIONS

Interpolation of $\overline{X}$ (in μ)				2.0	2.5		3.0			4.0						
t..........	0	5	10	15	20	25	30	35	40	45	50	55	60	65	70	75
x..........	75	70	65	60	55	50	45	40	35	30	25	20	15	10	5	0
y..........	0	10	20	30	40	50	60	70	80	90	100	110	120	130	140	150
z..........	225	220	215	210	205	200	195	190	185	180	175	170	165	160	155	150
Total.....	300	300	300	300	300	300	300	300	300	300	300	300	300	300	300	300
Hypothesis				7	6	5	4	3	2	1						
Limit.....	$y = 0$		Minimum probable value of y		Answer by interpolation				Minimum probable value of x							$x = o$

Actual cases will deviate from the ideal we assumed above. Hence we set the upper limit by letting one of the components have equal frequencies in all its classes, such as component B in HO 1, table 27. We set the lower limit by letting the same component have only one class, such as component B in HO 7, table 27. We know then the true solution will fall within these two limits. Estimations will then be limited to adjustments of frequency among the different classes between the different components until we arrive at a reasonable set of spectra representing the components.

Estimation can be greatly facilitated if a preliminary grouping in terms of morphological interpretations can be made.

Another method of approximation is shown in table 28. Use the maximum and minimum values of t as the extreme limits. When $y = 0$, $t = 0$. This is the minimum end. When $x = 0$, $t = 75$. This is the maximum end. This is obvious, for, when either x or y tends to zero, the mixture contains only two components. It is also obvious that the limits set by the t values have a range wider than we actually need.

The range between $t = 0$ and $t = 75$ is then subdivided into any number of columns. In this example the subdivisions were made at an increment of 5 (i.e., $t = 0, t = 5, t = 10, \cdots, t = 75$). Calculate for every column the values of x, y, z from the parametric equations. Plot the probable limits found by iteration shown in table 27 into table 28. These limits are hypotheses 1 and 7. Hypothesis 1 has $A = 30$, $B = 90$, and $C = 180$. It is exactly the same as x, y, and z under column $t = 45$ in table 28. Hypothesis 7 has $A = 60$, $B = 30$, and $C = 210$. This is the same column as under $t = 15$ in table 28. Interpolate maximum possible mean of the three components ($\overline{X} = 2$) under column $t = 15$, that is, where hypothesis 7 is located. Then, considering these limits as from 2 to 4, for an unknown sample with the mean 2.5, the column that gives the approximations will be between columns $t = 20$ and $t = 25$. The purely hypothetical result will be $x = 52.5$, $y = 45.0$, and $z = 202.5$. The correct answer that I set for the example was hypothesis 5 shown in table 27. The approximation comes very close to it.

Application to a Mixed Sample of Known *Pinus*

Samples of 100 grains each of *P. ponderosa* × *jeffreyi* 22B (A), *P. murrayana* 27B (B), and *P. ponderosa washoensis* 29B (C) (table 29) were drawn from a slide on which these pines were mixed in unknown proportions. In mounting the slide, one drop of a prepared sample of each species was used.

The mixed slide assumes that the universes of the components are equal and infinite, and it represents an unbiased sample of the three universes.

The sample drawn from this mixed universe is assumed to be a random sample. The individual components have also been studied separately, and a spectrum of each was obtained as reference. The reference composite spectrum, I, and its mean, $\overline{\overline{X}}_{\mathrm{I}}$, together with those of the unknown spectrum, II and $\overline{\overline{X}}_{\mathrm{II}}$, are given in table 29.

Inspection shows that $\overline{\overline{X}}_{\mathrm{I}}$ is larger than $\overline{\overline{X}}_{\mathrm{II}}$. This would indicate that the unknown is composed of a mixture of components A, B, and C in unequal proportions and that the one with a higher mean than average is lesser in quantity, and the one with a smaller mean than average must be more.

Substituting the individual means into equation (1), we have

$$76.49x + 58.40y + 65.48z = 19374, \qquad (6)$$

where x, y, and z represent the number of individuals of components A, B, and C, respectively, and their sum is 300 as in equation (3). Dividing both sides by the smallest mean, we reduced equation (6) to

$$1.31x + y + 1.12z = 331.747. \qquad (7)$$

Solving simultaneously (3) and (7), we obtain the limits of solutions between points when $x = 0$ and $y = 0$. Between these limits there are an infinite number of sets of three quantities which will satisfy equation (7).

In the sample drawn, preliminary grouping cannot be achieved by means of morphological considerations, so we must proceed directly by iteration.

The process of iteration is lengthy and time consuming. It is done by trial and error: by trying to fit the spectra of references into the unknown spectrum II. For each fitting we are trying to match the mean of the fitted individual spectrum with the mean of the particular reference sample on which the fitting is based. We also take care to see that the average mean of the fitted spectra matches that of the unknown spectrum II. Bearing in mind the pattern of frequency distribution in the individual components, we find but few tenable ways to fit the spectrum.

TABLE 29

Spectra of References and Hypothetical Fittings of Unknowns

Com-ponent	References	Pollen frequency by class midpoint																				$\overline{X}$ (or $\overline{\overline{X}}$ for total) (in μ)	N	C (in percent)
		38	41	44	47	50	53	56	59	62	65	68	71	74	77	80	83	86	89	92	95			
A	*P. ponderosa* X *jeffreyi* 22B									1	5	5	10	17	29	21	6	1	2	2	1	76.49	100	7.6
B	*P. murrayana* 27B	1	1	0	1	0	17	22	20	27	9	1	0	1								58.40	100	8.7
C	*P. ponderosa washoensis* 29B			1	0	1	3	5	6	11	29	23	12	6	2	1						65.48	100	8.9
	Total: spectrum I	1	1	1	1	1	20	27	26	39	43	29	22	24	31	22	6	1	2	2	1	66.79	300	
Unknown: spectrum II		6	0	3	6	6	30	21	12	42	54	36	18	15	33	9	9					64.58	300	
Fitting from hypothesis 5																								
A											2	2	4	13	27	8	9					76.59	65	5.4
B		6	0	3	6	6	24	19	7	26	21	12	6	2								58.39	138	13.7
C							6	2	5	16	31	22	8	0	6	1						65.34	97	8.4
	Total	6	0	3	6	6	30	21	12	42	54	36	18	15	33	9	9					64.58	300	
Fitting from hypothesis 6																								
A											2	3	3	15	27	8	9					76.47	67	5.2
B		6	0	3	6	6	24	19	10	26	21	14	7									58.41	142	13.5
C							6	2	2	16	31	19	8	0	6	1						65.46	91	8.5
	Total	6	0	3	6	6	30	21	12	42	54	36	18	15	33	9	9					64.58	300	
Fitting from hypothesis 8																								
A										1	1	1	8	11	31	9	9					76.49	71	5.4
B		6	0	2	5	5	29	16	7	30	20	15	2	2								58.40	139	13.2
C				1	1	1	1	5	5	11	33	20	8	2	2							64.73	90	8.5
	Total	6	0	3	6	6	30	21	12	42	54	36	18	15	33	9	9					64.58	300	

The fitting can be done in two ways: The first is to allot frequencies to the various classes, somewhat intuitively, for every component that is supposed to be there and calculate their means. If the average mean of these fitted spectra does not match the mean of the unknown spectrum, merely adjust the frequencies here and there until the average mean equals that of the unknown spectrum. Two results of this type of trial are shown in table 29, and hypothesis 5 gives a better approximation than 6. The second way is to try to match the individual means first, and if such is achieved, the average mean will automatically be equal to the mean of the unknown spectrum.

It is obvious that in a real case no knowledge of references will be given. These have to be determined by estimating what the components are, and estimating the identity of the components depends on the interpretation of the unknown spectrum. Studying the unknown spectrum will give ideas that lead at least to one of the unknowns and, from there on, the others will be found. With the references established, and their spectra and mean values determined, the problem is reduced to finding the quantity of each.

Iteration, then, is essentially a reasoning process.

FITTING OF *B*

From equation (7) we find that, when $z = 0$, $x = 102.42$ and $y = 197.58$, and that when $x = 0$, $z = 264.58$ and $y = 35.42$. These are the extremes. The in-between sets of values are shown in table 30. These are obtained from the parametric equations

$$x = 102.42 - t, \tag{8}$$

$$y = 197.58 - 1.5833t, \tag{9}$$

and

$$z = 31t/12 \text{ or } 2.5833t. \tag{10}$$

As we noted earlier, it is of advantage to fit in the spectra involving either tail of the unknown spectrum; those of *A* and *B* may be tried first, as *A* and *B* have the highest and lowest means. Here I deliberately chose that of *B* because it is more difficult and serves as a good illustration to the process.

The lower tail of *B* spectrum is governed by the frequencies 6 and 0 of classes 38 and 41, respectively (table 29, Unknown spectrum II). The mode appears at class 53 where the frequency is 30, but the frequencies of classes 56 and 62 are also high. Class 65 is believed to be the mode of the *C* spectrum. Assuming each component has 30 grains at its modes, then class 65 will have 24 grains left after the mode of *C* is deducted. A larger portion of these 24 grains belongs to *B*. From the distribution we may allot 16, or ⅔ of 24, to *B*, and 8 to *A*. It follows that class 62 of y should also contain higher frequencies. That y at 62 should be high is obvious from the fact that this class is the lower tail of *A* and would contain few grains, probably 1 or 2. Class 62 has 42 grains. The number in the spectrum of z should not be more than its mode, 30. We may assume it to be 25; then class 62 of y will have some 15 grains. The frequency in class 59 is low. It is situated halfway between the modes of *B* and *A* spectra, so we divide the number 12 equally and allot 6 to each. Class 56 has 21 grains, which number is located immediately next to the mode of *B*, and much farther from the mode of *C*, so we allot ⅔ (14

TABLE 30

VALUES OF t, x, y, AND z FROM DATA BY FINE GROUPING

$\overline{X}$ (in μ)	58.40	60.17	61.93		63.70	64.58	65.46		67.23	69.00	70.76	72.53	74.29	76.06	76.49
t..........	0	10	20	24.12	30	35	40	45.86	50	60	70	80	90	100	102.42
x..........	102.42	92.42	82.42		72.42		62.42		52.42	42.42	32.42	22.42	12.42	2.42	0
y..........	197.58	181.75	165.91		150.08		134.25		118.42	102.58	86.75	70.92	55.08	39.25	35.42
z..........	0	25.83	51.67		77.50		103.33		129.20	155.00	180.80	206.70	232.50	258.30	264.58
Limit......	$z = 0$														$x = 0$
Hypothesis.				HO 1 max (table 31)		Answer		HO 1 min (table 31)							

NOTE: The figure 64.58μ is the mean of unknown sample. The t values 24.12, 35(34.99), and 45.86 are interpolations. These are values for hypothesis 1, maximum and minimum, shown in table 31, and the value for the interpolated answer.

grains) to y and ⅓ (7 grains) to C. If B has taken up all 30 grains in class 53, that of C will be 0, and all the lower-end frequencies should belong to B only.

The estimations above indicate that B is a bimodal spectrum—one mode at 53, another at 65—and that the frequencies taper off rapidly after class 65. The numbers at classes 68, 71, and 74 would be low: probably something like 2, 1, 1. Now that we have a complete spectrum of B, we calculate its mean, and find that it is 55.85μ and that it contains 106 grains (i.e., $y = 106$). We know from references that this mean is lower than required (58.40μ, table 29, reference *P. murrayana*). We know also from references that spectrum for C should have grains in classes 44, 47, 50, and 53, to which we did not allot any. By reducing these classes in the B spectrum into 44 = 2, 47 = 5, 50 = 5, and 53 = 25, and by increasing the number in

TABLE 31

HYPOTHESES FOR SPECTRUM-FITTING OF x

Hypothesis	Class number	x	y	z	Pollen frequency by class midpoint								$\overline{X}$ (in μ)
					62	65	68	71	74	77	80	83	
Maximum short HO 1	5	77	157.3	65.7	..	..	..	13	14	32	9	9	76.494
Maximum long HO 2.	8	71	147.8	81.2	1	1	1	7	13	30	9	9	76.493
Minimum short HO 1	5	59	128.8	112.2	..	..	..	12	13	16	9	9	76.492
Minimum long HO 2.	8	47	109.8	143.2	1	1	1	7	9	10	9	9	76.490

classes 62 and above to 62 = 20, 65 = 16, 68 = 5, 71 = 2, 74 = 2, we obtained a second estimate. The result is that B now has only 102 grains and the mean is 57.06μ, which still falls short of our aim.

FITTING OF *A*

Since the fitting of B is complicated and has more classes to work on, and the results are not encouraging so far, it is therefore decided to try to fit the A spectrum first instead.

The quantity of A is limited by the upper tail of the unknown spectrum, which at class 83 has a frequency of 9 grains. Considering the mode at 77, the spectrum is either (a) very short or (b) skewed. It could be one containing only 5 classes ranging from class 71 to class 83, or one that extends down to class 62. The high frequency of class 65 indicates that within that class, all three components are present. Nevertheless we have to test both alternatives.

Hypothesis 1 table 31.—Considering only the five classes from 71 to 83, we must then have a spectrum roughtly symmetrical at class 77, and slightly unequal in number to account for the mean of 76.49μ instead of 77.0μ. We know (from table 29) that number of grains at class 83 is 9 and consider class 77 to be 30 grains, so class 80 in between these would be 9 at most. But, as class 80 is the tail of the C spectrum, it may be considered to be only 8 grains. Let classes 71 and 74 each have 10 grains; we will then have a spectrum for A of 10, 10, 30, 8, 9. This gives a total of 67 grains, and a mean of 76.84μ, which is slightly too high to fit the requirement.

The total number of grains in classes 71 and 74 is 18 and 15, respectively. If we allot the classes each with the maximum number permitted, that is 15 and 15, we have a spectrum of 15, 15, 30, 8, 9. This is a total of 77 grains, a mean of 76.25μ, which is a little too low.

Now we know the solution for x lies between $x = 67$ and $x = 77$ for its maximum. By repeated trials we find that a spectrum of 13, 14, 33, 9, 9 gives 77 grains with a mean of 76.482μ. A spectrum of 12, 15, 32, 8, 9 gives 76 grains and a mean of 76.487μ. A spectrum of 13, 14, 32, 9, 9 gives 77 grains and a mean of 76.494μ. The solution for maximum x with a short spectrum is therefore 77 grains. If $x = 77$, $y = 157.33$ and $z = 65.67$. giving an average mean of 65.48μ and a total of 300 grains, with all modes fitted correctly.

We have been estimating x for its maximum value. If the frequency at 77 mode should be the minimum, say the number is only slightly more than the 9 grains of classes 80 and 83, what estimation will we get? This we found to be $x = 59$, and a mean of 76.492μ.

Hypothesis 2 table 31.—There is, however, one fact that is of concern to us. The reference spectrum A (table 29) shows a range of 62 to 95. Since the unknown has no frequencies beyond class 83, and yet the mode is still at 77, we might reason that by chance those higher classes escaped sampling. By the same token, we reason that, for the classes lower than 71, frequencies, if present, would be very few in number. We have tested the hypothesis of short spectrum. We now have to try for the hypothesis that the distribution is skewed.

As before, we must find both maximum and minimum x. Let classes 62, 65, and 68 each consist of only 1 grain. Adding this portion to the short spectrum would pull down the mean value. These 3 grains added a total weight of 12 to the negative deviation. It is apparent that in order to restore the balance, these values have to be weighted by adding to the positive side. But we cannot do very much with the positive end, since nearly all available frequencies have been taken up. By loading it to the full capacity of 33, 9, and 9 we still need to reduce the number on the lower end in order to achieve the required mean. This we finally reach by a spectrum of 1, 1, 1, 7, 13, 30, 9, 9, which gives 71 grains and a mean of 76.493μ. Since $x = 71$, y will be 147.78 and z will be 81.17, totalling 299.95 grains. The minimum for x in a skewed pattern so far fitted would be 1, 1, 1, 7, 9, 10, 9, 9, which has 47 grains and a mean of 76.490μ; y will be 109.83 and z, 143.16; together with x, they make 299.99 grains.

REESTIMATION OF y AND ESTIMATION OF z

We must check B and C according to the indications given by the maximum and minimum values of x under *HO* 1 and *HO* 2 (table 31). It is now easy to see why we were not able to fit a suitable spectrum for B during our first trial, since the number of $B(=y)$ at a minimum should be 128 grains, while we were trying to solve it with 106 and 102 grains. Nevertheless, even with the indication given from x, the fitting of B and C spectra is still a lengthy process, for in between the limits of maxium and minimum values we have an infinite number of solutions. After a number of trials we arrived at the solution of hypothesis 5, shown in table 29. The solution gives $x = 65$, $y = 138$, and $z = 97$. As the mean values of each individual do not exactly fit the value given in the reference, the solution is at best an approximation.

Before discussing a more accurate solution, let us consider the parametric equations (8), (9), and (10). The change of y with respect to x is increasing by 1.5833 times, whereas the change of z is decreasing by 2.5833 times. A change of 1 in the number of x will therefore decrease the number of z by approximately 2.6 grains. Hence our estimation may be very close to the true x, but not as close to the true y, and less close to the true z. The rule of thumb would be that the estimated sample size of the member of the three components which has the intermediate mean value is liable to depart the most from the correct solution, or, in other words, be the least accurate.

CALCULATIONS AND INTERPOLATIONS

We may estimate the quantities of the components by calculating and interpolating. The values of t are arranged in a row, from $t = 0$ on the left to $t =$ maximum on the right (table 30). Over each of the t values is entered the corresponding value of mean, making the smallest mean of the three components coincide with $t = 0$, and the largest mean, with $t =$ maximum. Through calculation by parametric equations, the values of x, y, and z are found and are entered below every value of t. Calculate the t values from the x, y, z values given for maximum and minimum under hypothesis 1, table 31. Interpolate these t values in the table at their respective locations to scale. Drop a vertical line through each of these values. These are the limits within which the solution of the unknown will be located. Now interpolate along the row of mean values the mean of the unknown and also its corresponding t. Drop another vertical line. The frequencies of x, y, and z, along this line will be the frequency of components A, B, and C required for the unknown sample.

Calculation amounts simply to finding the t value of an unknown having a certain value of mean. The t is equal to mean of unknown, minus the smallest mean of the components, divided by the range of maximum and minimum means of the components, multiplied by t_{max}.

In our problem, $t_{max} =$ value of t when $x = 0$. From (8), $x = 102.42 - t$, then $t_{max} = 102.42$.

$$t \text{ for an unknown mean of } 64.58\mu = \left(\frac{64.58 - 58.40}{76.49 - 58.40}\right) (102.42)$$

$$= \frac{(6.18)\ (102.42)}{18.09}$$

$$= 34.99 \qquad (11)$$

Substituting $t = 34.99$ in (8), (9), and (10), we obtain

$$x = 102.42 - t = 67.43,$$

$$y = 197.58 - 1.5833t = 142.18,$$

$$z = 2.5833t = 90.39.$$

With these values as basis, iteration will yield the fittings shown under hypothesis 6, table 29. Note that the mean value of each component is much closer to the references than those given by hypothesis 5.

By using equation (11) we may also calculate the maximum and minimum values of t, by which we can obtain the maximum and minimum values for x. The standard error of 1.92μ is first found for the mean of the unknown. The confidence interval for the mean of 64.58μ will then be 66.50μ – 62.66μ. Substituting in turn these values in lieu of 64.58 in equation (11), we have

$$t_{\min} = \frac{62.66 - 58.40}{76.49 - 58.40}(102.42) = \frac{436.31}{18.09} = 24.12.$$

$$t_{\max} = \frac{66.50 - 58.40}{76.49 - 58.40}(102.42) = \frac{829.60}{18.09} = 45.86.$$

Substitute the values of t in equation (8). The maximum value of x will then be 78.30μ and the minimum will be 56.56μ, which are somewhat similar to the values we found through fitting under maximum and minimum short, *HO* 1, table 31.

The calculations derived by interpolation are much simpler than those derived by iteration. The fact that we tried the more difficult way first is designed to illustrate the usefulness of iteration in actual studies when no answer is given to check the results. Compare the results of hypothesis 5 with those of hypothesis 6 (table 29): we are in error by 2 grains for x, 4 grains for y, and 6 grains for z. The error is small considering we have three components with a difference of mean value up to 21μ.

FITTING BY COARSE GROUPING

In the fitting process, we have been handling a spectrum with a maximum of thirteen classes in y, which is very tedious and time consuming. A shorter method is to group the spectrum into coarser classes, thereby reducing the maximum number of classes per spectrum to six or seven.

In regrouping the data, the means of the individual components, the total spectrum, the parametric equations, and the t values must be recalculated (as shown in tables 32, 33).

The iteration process is much the same and we are able to achieve a better fit of $x = 68$, $y = 136$, and $z = 96$ (table 34). Hence, a coarser grouping seems to be of advantage in that it shortens the iteration process and achieves compatible results. Solving by interpolation yields $x = 67.42$, $y = 136.78$, and $z = 95.80$. We also found that the maximum for x is 86 and the minimum is 31.

CONCLUSIONS

1. Iteration is a dependable method for finding class number and quantities of the components of a mixed spectrum. The results from various different ways are comparable. Thus, the class number for x in hypothesis 5 is 65, in hypothesis 6 is 67, from interpolating is 67.4, and from coarse grouping, 68. The difference of three grains in the estimates is very small indeed. For y, we have 138, 142, 142, and 136, respectively, a maximum difference of only six grains.

2. The fitted spectrum has exactly the same frequency distribution as does

TABLE 32

DATA BY COARSE GROUPING

Component	Species and sample number	N	$\overline{X}_1$ (in μ)	Pollen frequency by classes in coarse grouping										
				35–38	41–44	47–50	53–56	59–62	65–68	71–74	77–80	83–86	89–92	95–98
A	*P. ponderosa* X *jeffreyi* 22B	100	77.54					1	10	27	50	7	4	1
B	*P. murrayana* 27B	100	59.42	1	1	1	39	47	10	1				
C	*P. ponderosa washoensis* 29B	100	66.02		1	1	8	17	52	18	3			
Total		300	202.98	1	2	2	47	65	72	46	53	67	4	1
Average			61.66											
Unknown spectrum		300	65.60	6	3	12	51	54	90	33	42	9		

TABLE 33

VALUES OF t, x, y AND z FROM DATA BY COARSE GROUPING

$\overline{X}_1$ (in μ)	59.42	612.3	63.04	64.86	66.68	68.48	70.29	72.10	73.92	75.73	77.34
t	0	10.23	20.46	30.69	40.92	51.15	61.38	71.61	81.84	92.07	102.32
x	102.32	92.07	81.84	71.61	61.38	51.15	40.92	30.69	20.46	10.23	0
z	197.70	179.85	162.00	144.13	126.30	108.45	90.60	72.75	54.90	37.05	19.25
y	0	28.09	56.16	84.24	112.32	140.40	168.48	196.56	224.64	252.72	280.90

NOTE: $x = 102.32 - t$; $y = 197.69 - 1.745t$; $z = 2.745t$.

TABLE 34

SPECTRUM-FITTING BY COARSE GROUPING

Samples	N	$\overline{X}_1$ (in μ)	Pollen frequency by classes in coarse grouping (upper row = midpoint of class groups)								
			36.5 35–38	42.5 41–44	48.5 47–50	54.5 53–56	60.5 59–62	66.5 65–68	72.5 71–74	78.5 77–80	84.5 83–86
Unknown	300	65.60	6	3	12	51	54	90	33	42	9
x	68	77.54	..	..	..	..	..	10	11	38	9
y	136	59.33	6	3	11	36	47	28	5	..	..
z	96	66.02	..	..	1	15	7	52	17	4	..
Maximum of x	86	77.52	..	..	..	..	..	2	33	42	9
Minimum of x	31	77.53	..	..	..	..	1	4	8	9	9

spectrum II and therefore the same mean. In some cases omitting fractions of a grain may result in an achieved mean slightly different from the required mean. So long as the difference does not exceed 2 per cent of that of the unknown sample, the result is acceptable.

IDENTIFICATION OF THE UNKNOWN

Preliminary Considerations

The discussion to now was based on the assumption that for some reason we knew the number of components, their identities, and their means. In this section, using the same sample as in the preceding section (table 29, spectrum II) and considering it as an unknown sample to begin with, I show how the number of components, their identities, and their respective frequencies can be estimated. This is a repetitive process.

First make an assumption as to the number of components. Second, try to formulate the span of the individual spectrum of these components. Third, from these spectra, calculate the statistics such as mean, class-C index, and sample number for each of the components. Identification of the components is made by comparison with reference data in appendix tables 1 and 2. Many trials may have to be made before getting close to the answer, and, for every new trial, the four-step process has to be repeated.

Given a spectrum such as spectrum II (table 29), we must, of course, decide at the beginning whether it is a spectrum of a mixed sample. Morphology is of little help here for no preliminary grouping is meaningful because all the pollen grains have psilate distal membrane and Y pattern on cappa. But it does assist in limiting our search to this particular group of pines. There are only a few pines with these morphological characters. They are *Pinus contorta* and its variety *bolanderi, P. murrayana. P. ponderosa* and its varieties, *P. jeffreyi* and its hybrids, and *P. coulteri*. Pollen grains of *P. attenuata* and *P. sabiniana* have both W and Y patterns. Since no grain of the W pattern was found among the unknown, its presence is therefore unlikely.

The class-C index of the sample is 16–15. The range is 38μ–83μ. There is no reference sample of any single species which matches the class-C index or the range. The sample must therefore be a mixed one. The number of components in a mixed sample with 16 classes must be either (1) two components very different from each other or (2) three or more components. For practical purposes, we started our fitting of the spectrum with three components. After the formulation of three spectra for the three components that we believe to be there, a test for homogeneity by the class-C indexes of the individual spectra is made. We accept the spectrum that passes the test. If any of these spectra show heterogeneity, it probably indicates that either we have more than three components or the entire series of spectra must be readjusted. If we can reasonably fit in three components with no surplus of frequencies and with no heterogeneity, there will be no fourth component. If we can identify the three components by the spectra formulated, there is no room for a fourth one either. What if there are only two components? This question is discussed below in the analysis of spectrum II.

MODAL DISTRIBUTION

Spectrum II (table 29) has three modes. They are at classes 65, 77, and 53. The frequencies of modes 77 and 53 are about the same weight and that of 65 is about twice the others. None of these modes has a frequency exceeding 20 per cent of the total sample size (= 300). Except for the lower tail, the distribution appears fairly symmetrical. Class 62 has a fairly high frequency, and class 56 should be regarded as a co-mode of 53, for their difference amounts only to 3 per cent.

From average mode frequencies and patterns shown in table 11 and figure 13, we derived the following conclusions:

1. If the three modes occur as trimodal, this can only be found among samples of *P. jeffreyi.*

2. If the three modes occur because of mixing of two samples of one taxon, then the component can only be *P. ponderosa.*

3. If the three modes occur as bimodal, then various combinations of two out of the three will indicate different taxa: for 65 and 53, *P. ponderosa;* for 65–62 and 53, *P. attenuata;* for 65 and 77, *P. sabiniana;* and for 53 and 62, *P. contorta* and *P. murrayana.* The spectrum is then composed of at least two components.

4. If the three modes occur as unimodal, then there are at least three components. For mode 53, *P. contorta, P. murrayana,* and their varieties are indicated. For mode 65. *P. ponderosa* is eligible; and for mode 77, *P. coulteri* is eligible. We also know that the unknown spectrum II, whether made up of two or three, or more, components, those components must be among the species listed here. Among them, we may eliminate *P. attenuata* and *P. sabiniana* because of morphological incompatibility. Hence we have left only *P. jeffreyi, P. ponderosa, P. contorta, P. murrayana,* and their varieties, and *P. coulteri.*

ANALYSIS OF SPECTRUM II

We begin with the component having a mode at class 65. Call it component *C* (table 29). The spectrum of *C* in the form we have it (table 29, hypothesis 5) is centered at class 65 and fairly symmetrical. Presumably its mean is very close to the mode and unimodal.

The quite regular rise of frequencies from two or three classes each side of class 65 toward the mode 65 indicates that this component is unimodal. From the general trend of the frequencies in spectrum II, the other two modes appear to belong to two other components with their modes at 53 and 77, respectively. If these also are unimodal, and if the modes coincide with the means, then for the three components we will have means 53, 65, and 77, and their average estimated mean is 65, not very different from the required mean of unknown spectrum II of 64.58μ. For a moment, the estimated mean looks satisfactory, but there lies in it an inconsistency. For, if the estimated means average about the same as the required mean, the components must be almost equally represented in the mixture. But we see from the spectrum II (table 29) that this cannot be so. Suppose component *A* is centered at 77, then classes 77 up to 83 will give a total frequency of 51 grains. In order for the mean to be 77μ, the frequency below the mode should be such that the sum of the negative deviation from mean cancels the sum of the positive deviation from the 18 grains in classes 80 and 83. In that case, the total number cannot be 100 and the components do not have equal quantities. If we make up three spectra, each containing 100 grains, we will also find that their means cannot be 53, 65, and 77, respectively. Hence the components must be unequal in quantity. It follows that the close approximation of the estimated mean, 65μ, to the required mean, 64.58μ, can indicate only that at least one of the components has a different mean from the ones we estimated. If component *C* is unimodal and its mode and mean should be 65, then any change in mean will be found in either component *A* or *B* or in both, and *A* and *B* may be bimodal.

Two possibilities present themselves:

1. Mode 77 is associated with either mode 65 or 62 in the component *A* spectrum, and the mean will then lie in between; and/or

2. Mode 53 is associated with some other mode (say, for instance, 62 or 65), forming a bimodal distribution that extends into class 68 with the mean lying in between. These modes will be in the spectrum for component *B*.

These possibilities may work very well. But when we turn to the references for comparison of range and mean, we find that no known species has a range from either 62 or 65 to 83, whereas there are samples of species showing a range from 38 to 68 or 71. This implies that the second possibility (for component B) can be true, but the first (for component *A*) cannot. Spectrum *A* is therefore not bimodal as we had postulated.

The spectrum for component *C* is automatically delineated while we are formulating the above possibilities for the other two components. That spectrum *C* has a range from 56μ to 74μ is substantiated by references. Hence we have on hand two eligible spectra: *B* and *C*. We need still to adjust spectrum *A*.

Since spectrum *A*, with mode at 77, is not bimodal, the spectrum appears skewed. We have reason to believe that the spectrum, if fully represented, extends into the 92 or 95 class. It is not shown in spectrum II merely because these high values escaped sampling. With one end of the spectrum not fully registered, the quantity of component *A* must be small, yet the sample mean remains high. In fact, component *A* will have the highest mean of all three. This realization fits in well with the scheme because the required mean, 64.58μ, is certainly smaller than the mean of component *A*, and the quantity of *A* must be less than a third of the total frequency in a three-component mixture.

Now we must adjust our estimations. Component *A* is considered unimodal. Its mean and mode are at class 77. Since 9 to 11 classes are the usual range for a spectrum with a high mean, such as that of component *A*, the range probably extends from 65 or 62 upward to the 92 or 95 class.

Component *B*, since it is bimodal with modes at 53 and somewhere among classes 62 to 68, where the frequencies are high, must have a mean between 57.5μ and 60.5μ. The spectrum shows five classes lower than mode 53. Neglecting the accidental mode of class 38 at end of a spectrum II, we estimate an extension of three classes above the second mode; then the range will be 38 to 68, or to 71, or even to 74.

Component *C*, with a mode and mean at 65, probably covers 9 classes from 53 to 77.

The average mean of $A + B + C$ in equal quantities will be 66.5μ or 67.5μ, notably higher than the required mean. This is in accord with the knowledge that they are unequal in quantity in the unknown. To fit the required mean of 64.58μ, some details must be adjusted. It appears that *B* must have the largest quantity and that *C* must have more grains than *A*, since *B* is the one with the smallest mean while *A* has the highest mean of all three. To adjust is to formulate spectra for these components until we arrive at a suitable set of three, each of which will simultaneously (1) fit the required mean, (2) account for the frequencies, and (3) match the identity, mean, and class-C index of a single species in the reference samples (appendix tables 1, 2).

I have not given all my adjusted estimations in tabulated form simply because there are many ways of formulating a spectrum for each component, many of which are eventually discarded because they do not fit the requirements. Of my trials I have given only three sets (table 29, hypotheses 5, 6, and 8). Each set arrived at is one that fits the requirements of the unknown. A final decision is to be made among these three.

SPECIFIC IDENTITIES AND QUANTITIES

Based on the preliminary analysis of the three components made in the preceding section, I proceed to find the eligible species and reference samples that will form the basis for fitting spectrum II.

From appendix table 2, we sort out the eligible reference samples that match the estimated range and mean of each component. For component *A*, its range being either from class 62 to class 95 or from class 65 to class 92, the eligibles are two *P. coulteri* samples and four *P. jeffreyi* samples. The estimated mean of component *A* is 77μ. Allowing an ample confidence interval of 3μ, we eliminate all samples that have mean above 80μ or below 74μ. This eliminates all samples of *P. coulteri* and leaves three of the four samples of *P. jeffreyi* as eligibles. Hence, component *A* is identified as *P. jeffreyi.* The samples of *P. jeffreyi* 12*N*, 27, and 22*B*.

For component *B*, having a range between classes 38 and 74 and a mean somewhere between 57.5 to 60.5μ, we have only two eligibles: *P. murrayana* 27*B* and *P. jeffreyi* 11. Since we have *P. jeffreyi* for component *A*, *P. murrayana* is therefore the identity for component *B* (for abnormality of *P. jeffreyi* 11, see Ting, 1965, p. 280).

For component *C*, having a range from class 53 to class 77 and a mean of 65μ, we have *P. murrayana* 12, *P. ponderosa* 16, *P. attenuata* 19, *P. sabiniana* 3*T* and 11. With the preliminary knowledge that the unknown sample contains no grains of the W pattern, *P. attenuata* and *P. sabiniana* are both out of the question. We have already chosen *P. murrayana* for component *B;* this leaves us with *P. ponderosa* as the identity of component *C*.

Knowing the identities of the components, we can now estimate the relative frequency of each which contributes to the unknown spectrum II. The samples for the components and their means are *P. jeffreyi* 27, mean 76.46μ; *P. jeffreyi* 22, mean 76.49μ; *P. murrayana* 27B, mean 58.40μ; *P. ponderosa* 16, mean 64.43μ. *P. jeffreyi* 27 is co-modal at classes 77–80, but *P. jeffreyi* 22*B* is unimodal at class 77; therefore we choose the latter sample for fitting of the spectrum.

The actual fitting is done in the same way as outlined in the section, Application to a Mixed Sample of Known *Pinus,* above. The results of fitting by modal analysis are entered, together with those previously obtained, in table 35, and the spectra of the three components are shown in table 29 as hypothesis 8. We see that the maximum discrepancy between estimations is only six grains.

TABLE 35

RESULTS OF FITTING SPECTRUM II BY VARIOUS METHODS

Component and quantity	Estimated pollen frequency of the three components in spectrum II					
	By fine grouping				By coarse grouping	
	By iteration		By modal analysis	By calculation and interpolation	By iteration	By calculation and interpolation
	Hypothesis 5	Hypothesis 6	Hypothesis 8			
A and *x*........	65	67	71	67.4	68	67.42
B and *y*........	138	142	139	142.2	136	136.78
C and *z*........	97	91	90	90.4	96	95.80
Total........	300	300	300	300.0	300	300.00

TABLE 36

SPECTRUM OF SACCUS-BREADTH MEASUREMENTS OF THE UNKNOWN POLLEN SAMPLE

N	$\overline{X}_2$ (in μ)	Pollen frequency by class midpoint														
		32	35	38	41	44	47	50	53	56	59	62	65	68	71	74
50..........	53.84	3	0	1	1	3	1	3	14	8	5	8	1	1	0	1

ANALYSIS OF THE SPECTRUM OF SACCUS MEASUREMENTS

We have so far employed only a few of the many statistics available and only one of the measurements we made on a pine pollen. We have used the measurement of the corpus, the mean, the mode, and the range, and also the class-C index. We have not used the measurement of the saccus, the overall size, the ratio, or the cos A index. The results are encouraging for the limited amount of information used. Experience has shown that measurements made of the corpus would be sufficient to identify components of an unknown. Nevertheless, we wish to try using other measurements such as those of the saccus, to see if we will arrive at equally encouraging estimates.

We are using the measurement of saccus breadth of the same unknown sample. Its spectrum is shown in table 36.

The spectrum of saccus has a range from 32μ to 74μ. It also shows three modes. Only one of these stands out prominently; the others are all low in frequency. But it does indicate that there are three components. Mode 44 correlates with *B*, mode 53 with *C*, and mode 62 with *A*.

We repeat what we did for the corpus. We first estimate the range and the mean for each component, and search through the tables of range and mode (appendix tables 1, 2) to find the eligibles and eliminate them by the mean value. We find a range of 32–56 for a nine-class spread of component *B* with a 44 mode. Only *P. murrayana* and *P. contorta* are indicated.

For component *C*, the estimated range will be 44–65; *P. ponderosa* 29 (50–65, 41–65, 44–65) is decidedly better than *P. attenuata* 12 (50–62) or *P. radiata* 6 (44–71). *P. ponderosa* 16 is not in the list of eligibles.

From the distribution of modes, component *A* is *P. jeffreyi* 22*B;* for component *B* we get no result; for component *C* we obtain a preponderance of *P. ponderosa* and *attenuata* 12.

A question may arise over our decision that *P. ponderosa* 29 is a better choice for component *C*. Apart from the morphological consideration in the measurement of corpus, we find that *P. ponderosa* 16 and *P. attenuata* 19 fit the range; *P. attenuata* 12, *P. ponderosa* 16, 29, 33 fit the modal distribution. *P. ponderosa* 16 is the only sample that fits both measures. In the saccus measurement, we find many *P. ponderosa* samples and only one *P. attenuata* sample (12) that fit the range and the modes. In each case, the *P. ponderosa* samples give a better fit than the *P. attenuata* samples. Hence we identify component *C* as *P. ponderosa.*

I have not included tables of spectra of saccus and of overall sizes in this work because I do not use them for identification of an unknown. This section on identification by saccus is included to show that we can also reach the same conclusion if we choose to use saccus measurements rather than corpus measurements. There is no conclusive evidence to say which measurement gives better results, although the estimate of quantities made by the saccus measurements in this particular case appears to be better (table 35, hypothesis 8).

NUMBER OF COMPONENTS

Now we come to the question of whether spectrum II is composed of fewer or more than three components. The estimates of quantity based on the assumption that there are three components are shown in table 35, and their spectra are shown in table 29. Each of these spectra has been tested by the class-C index. Except for the index for component *B* in hypothesis 5, all are normal indexes for single species. Since fittings of the spectra of both the corpus and the saccus leave no stray frequencies unaccounted for, the assumption of only three components and no more is valid and true. Incidentally, a consistent correlation between the three modes on the spectrum of the saccus and those on the spectrum of the corpus also strengthens our conviction of three components.

Is it possible to have only two components? We shall find that simultaneously to fit the three modes, the distribution of the frequencies, as well as the required mean with only two components from eligible reference samples is not possible. Briefly: take into consideration only the eligible species. *P. jeffreyi, P. ponderosa,* and *P. murrayana.* To fit spectrum II with only two components, one would have to be *P. jeffreyi,* and the other would have to be either of the other species. The eligible sample of *P. jeffreyi* in this case contains samples 2, 8, and 9*T,* each of which has a mean above 72μ. Every eligible sample of *P. ponderosa* and *P. mur-*

rayana has means lower than the required mean (64.58μ). Assuming equal proportions, to fit the required mean with *P. jeffreyi* 2, a mean of 75.08μ, we need a component with a mean of 54.08μ. If the quantity of *P. jeffreyi* is only one-third of the other components, the mean of the latter component should be 60.75μ. If the quantity of *P. jeffreyi* is three times that of the other component, the mean of the latter would have to be 33.08μ. It is evident that numerous combinations of mean and quantity will satisfy the required mean and quantity of spectrum II. Take these combinations of reference samples one by one and try to fit the spectrum. Only failure results. The fitted spectrum will not pass the class-C index test. If, on the other hand, we start by fitting the spectrum at random, we may arrive at two spectra perfectly normal as far as class-C index is concerned, but we will not be able to match them with any existing references. Hence we conclude that the unknown is composed of three components, no more, no less.

TABLE 37
STATISTICS OF ELIGIBLE SPECIES

Species and sample number	$\overline{X}_1$ (in μ)	$\overline{X}_3$ (in μ)	$\overline{X}_2$ (in μ)	Cos A	C (in percent)
P. jeffreyi X *ponderosa* 22B	76.49	102.56	62.24	6712	7.63
P. murrayana 27B	58.40	73.82	40.52	6153	8.70
P. ponderosa 16	64.43	93.56	56.69	73.68	7.76
Average	66.44	89.98	53.15	67.44	13.82

Ratios: $R_1 = 1.25004$; $R_4 = 1.35430$; $R_3 = 1.69483$.
Ratio index of mixed sample of known references: 1213.
Cos A index of average means: 6814.
Average cos A index of individuals: 6744.

Unknown	64.58	95.25	53.84	76.16	14.94

Ratio indexes: 1114, 1115, 1214, 1215.

RATIO INDEX AND COS A INDEX

It is not possible to use the ratio and cos A indexes of the unknown for purposes of estimation of quantity because the mean values of corpus, saccus, and overall size in the unknown are the result of differential mixing. The number measured for one component in the data of corpus may be different from that of the saccus or overall measurements of the same component. This is obvious from the fact that the unknown mean of overall size is larger than the reference average (table 37).

The numbers of individuals in the overall and saccus measurements can be calculated by the same method used for finding those of the corpus. We found that the numbers varied.

For ratio and cos A indexes, we must convert the individual mean values of saccus and overall size by assuming that they are sampled exactly as corpus size.

That is to say, each component has exactly the same amount of grains measured for saccus and overall sizes as that of the corpus measurements. For instance, if the numbers of *x, y,* and *z,* respectively, are, in round numbers, 67, 142, and 91, the values of $N\overline{X}$ for corpus, based on the reference data, will be (67) (76.49), (142) (58.40), and (91) (65.48). The sum of these three products divided by 300 will be 64.587μ, which scarcely differs from our required mean of 64.58μ. This will serve pretty well for estimating the $N\overline{X}$ values of overall and saccus, from which the ratios and cos A can be computed.

$$\text{For overall, } N\overline{X} = (67)\ (102.56) + (142)\ (73.82) + (91)\ (93.56)/300$$
$$= 25867.92/300$$
$$= 86.22640.$$

$$\text{For saccus, } N\overline{X} = (67)\ (62.24) + (142)\ (40.52) + (91)\ (53.72)/300$$
$$= 15082.71/300$$
$$= 50.2757.$$

$$R_1 = \frac{64.587}{49.373},\ R_3 = \frac{86.023}{49.373},\ R_4 = \frac{86.023}{64.587}$$
$$= 1.28465 \quad = 1.71507 \quad = 1.35504$$

$$\text{Cos A} = \frac{R_3{}^2 + 1 - R_1{}^2}{2R_3} = 0.6679.$$

The adjusted values are as follows:

Sample size: component $A: x = 67$

component $B: y = 142$

component $C: z = 92$

Mean of total sample $(A + B + C)$:

$\overline{X}_1 = 64.587\mu$

$\overline{X}_2 = 50.276\mu$

$\overline{X}_3 = 86.226\mu$

Ratios: $R_1 = 1.28465$, $R_3 = 1.71507$, $R_4 = 1.35504$

Ratio indexes: 1213, 1313

Cos A index: 6679.

We know that the ratio index of the mixed sample from known references is 1213 (table 37). Those of the unknown were originally 1114, 1214, 1115, and 1215 (table 37), which are totally different. But the adjusted data give ratio indexes of 1213 and 1313, which are compatible with that of references. This by itself confirms our estimation in hypothesis 6, table 35.

We can also tell that we are correct from the adjusted cos A index. The cos A index of the three known components averages 6744; that of the unknown through adjustment is 6679, a difference of 65. The small difference also confirms the correctness of the estimation, for the maximum difference between two lots within one universe is 70 (see appendix table 6).

TABLE 38

EXAMPLE I. SPECTRA AND STATISTICS OF LOTS AND TOTAL SAMPLE OF FOSSIL SAMPLE; MERCED 2163
(Modes in italics)

Lot	Pollen frequency by class midpoint																				N	$\overline{X}_1$	$\overline{X}_2$	Ratio	2S E	Class-C index
	32	35	38	41	44	47	50	53	56	59	62	65	68	71	74	77	80	83	86	89		(in μ)				
A Sublot 1	..	..	1	1	2	2	1	*9*	4	4	3	..	..	..	..	..	..	..	..	..	27	53.11	.....	......	....	9–12
A Sublot 2	..	..	..	..	..	..	..	..	..	..	..	1	6	*9*	2	2	1	2			23	72.17	.....	......	....	7–7
A total	..	..	1	1	2	2	1	9	4	4	3	1	6	9	2	2	1	2			50	61.88	84.02	1.3578	3.15	16–18
B	..	2	2	5	6	6	*12*	*14*	9	13	*25*	9	9	*24*	6	4	*11*	5	0	1	163	61.94	87.34	1.4100	1.83	19–19
A + *B*	..	2	3	6	8	8	13	*23*	13	17	*28*	10	15	*33*	8	6	*12*	7	0	1	213	61.93	86.56	1.3977	1.58	19–19
C	2	2	6	4	3	3	2	*9*	3	4	4	4	1	2	0	0	1	..	..	..	50	51.56	.....	......	3.23	17–22
Total	2	4	9	10	11	11	15	*32*	16	21	*32*	14	16	*35*	8	6	*13*	7	0	1	263	59.96	.....	......	1.49	19–20

Example I. Fossil Sample 1, Merced 2163

Fossil sample 1, Merced 2163, is from the upper part of the Lower Merced formation in California and is of Late Pleistocene age. The pollen grains were extracted from the sediments by hydrofluoric acid treatment and acetolysis, and separation was by flotation in bromoform (specific gravity 2.0).

PRELIMINARY ANALYSIS OF THE SPECTRA

The pollen grains have psilate distal membrane and are therefore all from the Group B pines. Both Y and W patterns are present.

Three lots, *A, B, C* (table 38), were drawn from three different slides. Their spectra, means, and class-C indexes are shown in table 38. Lot *A* was arbitrarily divided into two sublots simply by recording grain sizes smaller than 63.5μ in sublot 1 and those above 63.5μ in sublot 2. The spectrum of sublot 1, the mean, and the class-C index are comparable to many samples of *P. contorta* and *P. murrayana,* but to none of the other species. The spectrum, mean, and class-C index of sublot 2, however, found no match in our reference samples.

Lot *A* and lot *B* are in fact identical in their means of corpus breadth and not significantly different from each other in their means of overall breadth. Both class-C indexes are high. The lots therefore are both heterogeneous. They are polled into one sample: *A* + *B*.

Lot *C* has only 50 grains, but the class-C index is abnormal. This may be the result of mixing of many grains of a low mean (i.e., *P. contorta* and *P.c.* var. *bolanderi*) with only a few grains of other species with high means scattered at the upper end of the spectrum.

The total sample, *A, B,* and *C,* consists of 263 grains and has a mean of 59.96μ and a class-C index of 19–20. The spectrum of the total sample, ranging from class 32 to class 89, has four modes. Three of these are about equal in frequency. Mode at class 71, with 35 grains, is the major. Modes at classes 62 and 53, each with 32 grains, and mode at class 80, with 13 grains, are the minors.

In tables 39 and 40 are listed all species and sample numbers of occurrences of these modes, in various patterns, which can be found from the spectra of references (appendix table 2). It can be seen from the table that the major mode at class 71 is much better fitted by the spectrum of *P. radiata* 1 than by that of *P. ponderosa* 12, the only two eligibles for a major mode at class 71. For mode at 80, *P. jeffreyi* and *P. coulteri* samples appear eligible, but these have their major mode at class 80. If either of these two is indeed one of the components, its quantity must be relatively small, and scattered in the classes at the upper end of the spectrum.

For modes at 62 and 53 there are numerous possibilities. These modes either occur as two unimodal patterns or as a number of bimodal patterns. The trimodal pattern found in connection with *P. attenuata* 7 is favorable (table 39, row 10), for it not only gives us the fitting of modes at 62 and 53, but it also adds some quantities to the mode at 71, which in the fossil spectrum is the major mode. Since mode 53 in the spectrum of *P. attenuata* 7 is the smallest of the three modes, we therefore would need a bimodal or even a unimodal component with major mode

TABLE 39

SPECIES AND SAMPLE NUMBER OF REFERENCE SAMPLES WITH ONE OF THE MODES OF THE FOSSIL SPECTRUM AS MAJOR MODE

Row	Modal distribution Modes that occur in a spectrum		Species and sample number of reference samples under specific pattern			
	Major mode	Minor mode	Unimodal	Bimodal	Trimodal	Multimodal
1	80	71, 86	*P. jeffreyi* 27	*P. coulteri* 6	*P. coulteri* 11	
2	80–77		*P. jeffreyi* 12			
3	80–74			*P. ponderosa* 27		
4	71	62–59, 77–80–83	*P. ponderosa* 12		*P. radiata* 1	
5	71	62, 77, 53–56				*P. sabiniana* 20
6	71	65–62		*P. attenuata* 25 *P. ponderosa* 3, 41		
7	71–68–65	77	*P. ponderosa* 12N	*P. radiata* 3A		
8	62	53	*P. murrayana* 2, 12, 17, 23, 27B *P. ponderosa* 6, 8, 24T, 31, 35 *P. remorata* 4 *P. muricata* 12	*P. ponderosa* 5 *P. attenuata* 2, 11, 13, 16, 19, 20, 30, 30T *P. muricata* 1B *P. remorata* 3 *P. radiata* 7 *P. sabiniana* 2B		
9	62–65	53	*P. sabiniana* 3	*P. ponderosa* 1, 19		
10	62–65	71, 53–56			*P. sabiniana* 10 *P. attenuata* 7	
11	62	56, 80		*P. ponderosa* 15 *P. attenuata* 14	*P. sabiniana* 7	
12	62	77		*P. sabiniana* 9		
13	62–65–68	50	*P. ponderosa* 37	*P. coulteri* 8		
14	53	62, 68	*P. contorta* 1, 6, 11, 22A, 24, 27 *P. c.* var. *bolanderi* 3, 5, 6 *P. c.* var. *latifolia* 8, 23, 26 *P. murrayana* 1, 4, 5, 8, 20, 22, 27A, 34, 37 *P. muricata* 1B, 2 *P. remorata* 5	*P. contorta* 7A, 17, 22, 22B, 25 *P. murrayana* 10, 19, 28, 30 *P. attenuata* 3, 24 *P. muricata* 9, 10	*P. muricata* 7	
15	53–56	62		*P. contorta* 15, 18 *P. murrayana* 15 *P. radiata* 2		

at 53 in order to maintain the proportion between these two modes in the fossil spectrum. The eligibles given in table 40 center very heavily on *P. contorta* and *P. murrayana* (table 39, rows 14, 15). As I mentioned earlier, sublot 1 could possibly be fitted by *P. contorta* or *P. murrayana,* and lot *C* could have *P. contorta* and its variety *bolanderi* as its components. The preponderant existence of samples from these species over others having the mode at 53 strengthens my previous assumptions.

TABLE 40

SPECIES AND SAMPLE NUMBER OF REFERENCE SAMPLES WITH ONE OR MORE OF THE MODES OF THE FOSSIL SPECTRUM AS MINOR MODES ONLY

Row	Modal distribution		Species and sample number of reference samples under specific pattern			
	Major mode	Minor mode	Unimodal	Bimodal	Trimodal	Multimodal
1	77	71		*P. sabiniana* 18		
2	68–65	80, 56, 50				*P. ponderosa* 7
3	68	62–59		*P. coulteri* 2A		
4	68	62		*P. ponderosa* 40		
5	68	53		*P. attenuata* 29		
6	65	74–71, 53		*P. ponderosa* 3	*P. jeffreyi* 20	
7	65	53		*P. ponderosa* 22 *P. attenuata* 21		
8	65	56, 71, 77			*P. sabiniana* 3T	
9	59	53		*P. contorta* 2		
10	59–62		*P. attenuata* 30T *P. radiata* 3B			

We obtained some idea of what the components are from this analysis of the modal pattern of the fossil spectrum. There are probably three dominant components and a couple of minor components. *P. radiata, P. attenuata,* and *P. contorta* (or its variety *bolanderi,* or *P. murrayana*) will be the dominants.

In table 39, the listed samples are only those that happened to have one of the four modes in the fossil spectrum as the major mode. In a mixture there are certainly cases where modes are created out of an aggregate of a number of component spectra at one class while at the same time, literally speaking, "smothering" the modes that prevail in some other classes. The frequency distribution at the lower end of the fossil spectrum from classes 38 to 47, consisting of about the same quantities, could very well be the result of overlapping of a bimodal component beginning at class 41, having a mode at class 53, and a component that has a unimodal distribution centering on classes 41 and 44. Similarily, the frequency of class 59 in the fossil spectrum could be the result of similar overlapping of two bimodal spectra with 62, 53, and 59, 53, modes respectively. There is no way to know what the components are unless we go through a trial-and-error process. A logical start is to consider the modal distribution of the fossil spectrum as basis

TABLE 41

Data of Eligible Species and Samples Used in Spectrum Fitting of Fossil Sample, Merced 2163

Species and sample number	$\overline{X}_1$ (in μ)	N	$\overline{X}_2$ (in μ)	Pollen frequency by class midpoint (in μ)																							
				32	35	38	41	44	47	50	53	56	59	62	65	68	71	74	77	80	83	86	89	92	95	98	101
P. attenuata 19	63.89	100	86.84	..	..	..	..	..	..	..	6	9	6	31	17	14	9	7	..	..	..	..	..	..	..	..	..
P. attenuata 14	59.48	100	87.62	..	..	..	1	0	1	4	13	20	10	28	14	8	1	..	..	..	..	..	..	..	..	..	..
P. attenuata 24	56.96	100	80.99	..	..	1	3	0	2	4	27	14	18	21	7	2	1	..	..	..	..	..	..	..	..	..	..
P. attenuata 30	61.64	100	109.80	..	..	..	..	..	2	4	10	8	10	30	18	10	4	2	0	2	..	..	..	..	..	..	..
P. attenuata 11	61.91	100	87.83	..	..	..	..	..	..	3	6	7	10	32	31	7	2	..	..	..	..	..	..	..	..	..	..
P. contorta var. *bolanderi* 3	51.89	100	65.96	..	..	1	2	8	7	23	37	10	4	5	..	..	..	..	..	..	..	..	..	..	..	..	..
P. contorta var. *bolanderi* 6	52.46	100	69.44	..	..	..	6	0	2	26	48	6	6	6	..	..	..	..	..	..	..	..	..	..	..	..	..
P. contorta 1	52.43	100	70.61	..	..	..	..	1	9	28	39	16	7	..	..	..	..	..	..	..	..	..	..	..	..	..	..
P. contorta 7A	57.98	100	80.45	..	..	..	1	1	0	3	23	20	18	21	8	4	1	..	..	..	..	..	..	..	..	..	..
P. contorta 11	52.82	50	74.90	..	..	..	1	0	2	10	26	9	1	0	1	..	..	..	..	..	..	..	..	..	..	..	..
P. contorta 17	56.24	50	78.44	..	..	..	..	..	3	3	14	12	6	10	1	..	..	..	..	..	..	..	..	..	..	..	..
P. contorta 22	57.11	105	83.48	..	..	..	..	1	1	7	37	13	14	17	11	4	..	..	..	..	..	..	..	..	..	..	..
P. contorta var. *lectifolia* 23	50.18	100	64.34	..	..	2	6	8	16	24	32	6	4	2	..	..	..	..	..	..	..	..	..	..	..	..	..
P. contorta 24	51.74	50	74.36	..	..	1	1	3	4	9	24	4	2	2	..	..	..	..	..	..	..	..	..	..	..	..	..
P. contorta 25	54.68	50	72.86	..	..	..	1	1	1	5	26	4	2	6	4	..	..	..	..	..	..	..	..	..	..	..	..
P. jeffreyi 12	76.10	100	109.76	..	..	..	..	..	..	..	..	..	1	3	6	7	13	11	20	21	8	5	4	1	..	..	..
P. coulteri 6	75.68	50	103.70	..	..	..	..	..	..	..	1	0	0	2	1	5	8	6	4	15	4	3	0	1	..	..	..
P. muricata 12	63.94	100	95.10	..	..	..	..	..	..	..	3	7	14	27	22	15	4	3	1	4	..	..	..	..	..	..	..
P. ponderosa 1	62.27	100	84.83	..	..	..	..	..	..	4	11	7	10	24	23	12	4	2	3	..	..	..	..	..	..	..	..
P. ponderosa 7	68.24	100	94.22	..	..	..	..	1	1	5	2	5	3	5	17	18	12	11	6	9	1	2	2	..	..	..	..
P. ponderosa 15	61.52	100	91.79	..	..	..	..	1	1	0	9	10	7	36	30	1	3	2	..	..	..	..	..	..	..	..	..
P. ponderosa 31	60.56	100	90.08	..	..	..	..	..	2	2	6	12	14	44	14	6	..	..	..	..	..	..	..	..	..	..	..
P. ponderosa 27	71.48	50	95.35	..	..	..	..	..	..	..	..	2	3	2	5	6	7	10	4	10	1	..	..	..	..	..	..
P. radiata 1	65.63	100	85.42	..	..	..	1	1	1	2	10	10	10	12	8	5	15	3	8	7	6	1	..	..	..	..	..
P. radiata 3A	67.88	50	......	..	..	..	..	..	..	..	..	2	4	7	8	10	10	1	5	2	1	..	..	..	..	..	..
P. radiata 6	64.58	50	90.04	..	..	..	..	1	0	1	3	2	6	6	12	9	5	4	1	..	..	..	..	..	..	..	..
P. radiata 7	60.20	100	88.16	..	..	..	..	2	1	0	19	13	12	22	18	7	4	2	..	..	..	..	..	..	..	..	..
P. remorata 6	58.00	100	76.61	..	..	..	..	..	2	6	20	16	22	24	5	5	..	..	..	..	..	..	..	..	..	..	..
P. sabiniana 2D	68.33	100	94.94	..	..	..	..	2	0	1	4	4	1	18	3	17	29	3	9	5	1	0	2	0	0	1	
P. sabiniana 7	65.42	100	95.26	..	..	..	..	..	..	2	0	7	3	12	8	3	4	4	2	5	..	..	..	..	..	..	..

for iteration. Components that may have modes other than those of the fossil spectrum will reveal themselves during the iteration process.

In table 40 are listed the species and sample number of reference samples having one or more of the modes of the fossil spectrum as minor modes. The two tables, 39 and 40, give a key to the many samples we may use in fitting the fossil spectrum. No doubt after elimination by range and mean, many of these would not be useful at all.

FITTING OF SPECTRUM $A+B$: FIRST TRIAL

The first trial resulted in failure. I include it only to show the process of fitting a spectrum. The practice usually begins with breaking the original fossil spectrum into a number of hypothetical, but apparently homogeneous, spectra with the hope that these will match the modal pattern and distribution and mean of reference spectra of single species (appendix table 2). Spectrum $A + B$ of Merced 2163 is used as a beginning. The data on this spectrum are as follows (see also table 38): $\overline{X}_1 = 61.93\mu$; $\overline{X}_2 = 86.560\mu$; $R_4 = 1.3977$; modes are 71, 62, 53, 80.

From our modal analysis above and by scanning the spectra in table 41, the likelihood of a fit by *P. attenuata* and *P. contorta* is apparent. Only a few of the species and samples listed in table 39, however, have a mode at 80. Among them *P. coulteri* seems to be better than others. It appears better because *P. jeffreyi* 12, which has a mode at 80, also has an almost equivalent frequency at the 77 class. *P. ponderosa* 15 and *P. sabiniana* 7 also have an 80 mode but they will not give a proportionally high 71 mode to fit the $A + B$ spectrum.

The drawback in *P. coulteri* is its high mean (75.68μ). In order to counterbalance it, a species with very small mean, as small as 48μ, is necessary, and this is lacking in the list of eligibles. Take the sample with a very small mean: *P. contorta* 24, mean $= 51.74\mu$. We find that with a combination of 85 per cent *P. coulteri* and 115 per cent *P. contorta* 24 we are able to match the sample mean of 61.925μ. But we also find that the spectrum resulting from this calculation has very high frequencies at classes 53 and 80, which is contrary to the modal distribution of the unknown.

The following shows how the above conclusion is reached:

Unknown mean $= 61.93\mu$; $\overline{X}_1$ of *P. coulteri* 6 $= 75.68\mu$; therefore the required mean of the second component $= (2)(61.93\mu) - (75.68\mu) = 48.16\mu$.

The required species must have a $\overline{X}_1 = 48.16\mu$. The $\overline{X}_1$ of *P. contorta* 24 $= 51.74\mu$, which differs from the required by 3.58μ.

Assuming this difference arises from unequal quantities of each component, using equations (1) and (3), and modifying it to suit only two components instead of three, we have

$$Ax + By = (2)(61.93)(100), \text{ where } x \text{ and } y \text{ are unknowns.}$$

and

$$A = 75.68 \text{ and } B = 51.74.$$

Substituting, we have

$$75.68x + 51.74y = (2)(61.93)(100) = 12386, \qquad (12)$$

and

$$x + y = 200. \qquad (13)$$

Solving equations (12) and (13) simultaneously, we obtain

$$x = 85 \text{ and } y = 115.$$

This means that N of component A (*P. coulteri* 6) = 85 grains, and N of component B (*P. contorta* 24) = 115 grains. The modal distribution of unknown $A + B$ is 71, 62, 53, 80. The modal distribution found from the components, *P. coulteri* 6 and *P. contorta* 24, is 53, 80, 71, 62. The answer does not fit the requirement. Hence we arrive at the conclusion that a mixture of only two components cannot fit the unknown spectrum of Merced 2163, with *P. coulteri* as one of them. This shows that such a combination is not the complete answer. Other components are needed to fill in the spectrum in order to maintain the high frequencies at the 71 and 62 modes. Since by previous analysis the 62 and 71 modes do not associate, except in a special way in *P. attenuata* 7 and 25 (table 39, rows 6, 10), the number of components we need in addition to *P. coulteri* and *P. contorta* is at least two.

If there were four components in the spectrum instead of two, considering both *P. coulteri* and *P. contorta* to be two of the known constituents, then the others together must have a mean of 58.355μ (table 42). Then, *P. attenuata* 14, *P. remorata* 3, and *P. contorta* 7A can all fit the description. But we already have *P. contorta,* so the choice should be from the other two. *P. attenuata* 14 has a high mode at class 56 which is lacking in the original spectrum and which therefore has to be adjusted (see table 39, row 12). It will take a considerable quantity of *P. remorata* 3 (table 39, row 8) to bolster the mode of the 62 class, but the class 53 mode of *P. remorata* will simultaneously jump to larger proportions. Thus, balancing the modes of classes 71 and 80 would be impossible unless we put in more grains of *P. coulteri.* This goes around in circles.

It appears that the inclusion of some other species is necessary to strengthen the mode at 71 which will bring the list of species to five. A species preferably with high class 62 and class 71 frequencies and low class 53 and class 80 frequencies would be the ideal fit. We find from the listings that *P. radiata* 1 and *P. sabiniana* 20 best answer our need. So we try to fit the spectrum with five species: *P. coulteri, P. contorta, P. remorata, P. sabiniana,* and *P. radiata.* The fitting can be worked out, but the resultant mean value is too high. The frequency of mode 53 is too high. In reducing the frequency of class 53 by removing *P. contorta* or *P. remorata,* the substitution would have to be of a species with high mean value, and this would result in an even higher mean. Adjustment by varying the sample number may depress the mean, but it prevents reaching the modal requirements. The conclusion is, therefore, that this entire attempt is not successful.

If we decide that *P. coulteri* is the basic component and try to find the others, say two to begin with, then we may work it out as follows. If the mean of corpus of *P. coulteri,* plus the means of two other components, divided by three, equals the sample mean, then that sum of the means of the other two components must be 110.095μ. The mean for the overall size would then be 155.994μ. Their ratio will be 1.417. The various possible combinations of two components from known references are shown in table 42.

In the lower part of table 42 are listed eleven combinations of two components, their sum of $\overline{X}_1$ and $\overline{X}_3$, and the R_4 of the two means. The best fit to the mean will be the combination having the least difference of its R_4 from the one required (R_4 of unknown = 1.417). It is obvious that combination 3 is the best. The difference between the ratios amounts to only 0.003. Also good are combinations 2, 4, and 5.

TABLE 42

STATISTICS OF COMBINATION OF THREE COMPONENTS AND METHOD OF DETERMINING THE OTHERS WHEN ONE COMPONENT IS KNOWN

Method	$\overline{X}_1$ (in μ)	$\overline{X}_3$ (in μ)	R_4
Unknown sample *A* + *B*	61.925	86.560	1.398
Multiplied by 3	= 185.775	259.680	
P. coulteri 6	(−) 75.680	103.707	
Remainder = sum of two other unknowns	110.095	155.973	1.417
Let *P. contorta* 24 be one	(−) 51.740	74.360	
Remainder = the other unknown	58.355	81.613	1.398

Use the remainder to find the identity of the components by matching the values of the remainder with reference samples in appendix table 1.

Combination	Species and sample number of components	$\overline{X}_1$ (in μ)	$\overline{X}_3$ (in μ)	R_4	Difference of ratio value from unknown
	Required statistics of two unknowns combined	110.095	155.973	1.417	0
1	*P. contorta* 24 + *P. attenuata* 14	111.22	161.98	1.4560	.039
2	*P. contorta* 24 + *P. attenuata* 24	108.70	155.35	1.429	.012
3	*P. contorta* 11 + *P. attenuata* 24	109.78	155.89	1.420	.003
4	*P. contorta* 22 + *P. bolanderi* 6	109.57	152.92	1.396	.021
5	*P. contorta* 23 + *P. ponderosa* 31	110.74	154.42	1.394	.023
6	*P. contorta* 23 + *P. attenuata* 14	109.66	151.96	1.386	.031
7	*P. contorta* 23 + *P. radiata* 7	110.38	152.50	1.382	.035
8	*P. bolanderi* 3 + *P. attenuata* 14	111.37	153.58	1.379	.038
9	*P. contorta* 25 + *P. attenuata* 24	111.64	153.85	1.378	.039
10	*P. contorta* 7A + *P. bolanderi* 6	110.44	149.89	1.357	.060
11	*P. bolanderi* 3 + *P. attenuata* 24	108.85	146.95	1.350	.067

The spectra of these combinations are then put together with that of *P. coulteri* 6 and summed. We find, then, that a combination of *P. coulteri* 6, *P. contorta* 22, and *P. bolanderi* 6 gives too high a frequency at mode 53, while the combination of *P. coulteri* 6, *P. contorta* 23, and *P. ponderosa* 31 gives too low a frequency at mode 71. The other combinations give even poorer fittings. The fitting by three components consistently resulted in a high mode at class 53 and far too low a mode at class 71. I therefore hope that introducing one or two other components into the picture will compensate for the weakness.

TABLE 43

EXAMPLES OF UNSUCCESSFUL TRIALS IN SPECTRUM-FITTING OF FOSSIL SAMPLE MERCED 2163 WITH *P. jeffreyi* AND *P. coulteri*

Samples	Pollen frequency by class midpoint																				N	$\overline{X}_1$ (in μ)
	35	38	41	44	47	50	53	56	59	62	65	68	71	74	77	80	83	86	89	92		
Unknown	2	3	6	8	8	13	23	13	17	28	10	15	33	8	6	12	7	0	1	..	...	61.93
Hypothesis J2																						
P. jeffreyi 12	..	..	..	..	..	..	..	..	1	3	6	7	13	11	20	21	8	5	4	1	100	76.10
P. contorta 24	..	1	1	3	4	9	24	4	2	2	..	..	..	..	..	..	..	..	..	..	50	51.74
P. sabiniana 7	..	..	..	..	..	2	0	7	3	12	8	3	4	4	2	5	..	..	..	..	100	65.42
P. attenuata 19	..	..	..	..	..	..	6	9	6	31	17	14	9	7	1	..	..	..	..	..	100	63.89
P. radiata 1	..	..	1	1	1	2	10	10	10	12	8	5	15	3	8	7	6	..	..	..	100	65.63
Total	..	1	2	4	5	13	40	30	22	60	39	29	41	25	31	33	14	5	4	1	450	65.98
Hypothesis C2																						
P. coulteri 6	..	..	..	..	..	..	1	0	0	2	1	5	8	6	4	15	4	3	0	1	100	75.68
P. contorta 11	..	..	2	0	4	20	52	18	2	0	2	..	..	..	..	..	..	..	..	..	100	52.82
P. attenuata 24	..	1	3	0	2	4	27	14	18	21	7	2	1	..	..	..	..	..	..	..	100	56.96
P. radiata 7	..	..	..	2	1	0	19	13	12	22	18	7	4	2	..	..	..	..	..	..	100	60.20
P. ponderosa	..	..	..	..	..	4	11	7	10	24	23	12	4	2	3	..	..	..	..	..	100	62.27
Total	..	1	5	2	7	28	110	52	42	69	51	26	17	10	7	15	4	3	0	1	500	61.59

Looking through the spectra listed in table 41, it it appears to me that *P. sabiniana* 20 and *P. radiata* 1 are both capable of increasing frequencies at class 71. The inclusion of either one of them will result in a much higher mean than the required. Hence this particular attempt is not useful.

The above has been tried with *P. coulteri* 6 as one of the components. It suffices to say that I also tried with *P. jeffreyi* 12 (table 43, hypothesis *J*2 in lieu of *P. coulteri* 6 (table 43, hypothesis *C*2), and many such trials were made. Combinations that correctly fitted the modal distribution always have a higher mean than required, while correct fits to the mean value are always associated with a misfit in the relative frequency of the modes. It is impossible to achieve both with either *P. coulteri* or *P. jeffreyi* as one of the components. Hence these species were finally discarded as possible components. The fittings by the two species, *P. coulteri* 6 and *P. jeffreyi* 12, are shown in table 43.

FITTING OF SPECTRUM $A+B$: SECOND TRIAL

Having decided to abandon *P. coulteri* and *P. jeffreyi* as possible components, the remaining species with a high mean value are *P. sabiniana* and *P. radiata*. *P. ponderosa* is also a possibility. Tables 39 and 40 give the sample numbers of these species.

Reexamining the spectrum $A + B$, I find that the mode at class 80 is produced by grains occurring in spectrum *B* but not in *A* (table 38, lot *A* total, *B*). Spectrum *B* has a mode at 62, while spectrum *A* does not. The approach to the problem, then, is to keep spectra *A* and *B* separate and see if we can fit them individually.

For spectrum *B*, which contains 163 grains, I reach, after trial and error, a fitting that contains the following (table 44, hypothesis 3):

P. contorta 4	70 grains
P. radiata 1	100 grains
P. attenuata 30	100 grains
P. ponderosa 7	50 grains
P. sabiniana 20	50 grains

The total of 370 gains gives a mean of 61.934μ, only 0.011μ lower than that required (table 38, $\overline{X}_1$ of $B = 61.945\mu$; also table 44). The fit is good in many respects. Not only have all the modes been fitted but the relative frequency among them is more or less maintained. The mode at class 53 is slightly too high, but the fitting gives an acceptable mode at class 80, which has been the stumbling block in the previous trials. In attempting to adjust the frequency of mode 53, as well as the slight difference of the mean, I arrived at an adjusted fitting (table 44, hypothesis 4). The changes made are in the quantity of *P. contorta,* from 80 to 70 grains, and in that of *P. sabiniana,* from 50 to 71 grains. The mean will then be 61.927μ, differing now by 0.018μ from the required mean of 61.945μ.

Having settled on the components and quantity of *B,* we now try to find those of *A*. I formulate a number of subspectra for *A* as shown in table 45. These are *Aa, Ab* (*Ab*1 and *Ab*2), whose statistics are shown in table 46.

Apparently these spectra are homogeneous and represent two species. For *Aa,*

TABLE 44

Fossil Sample Merced 2163: Fitting of Spectrum *B*, Hypotheses 3 and 4, Second Trial

Sample	Pollen frequency by class midpoint																						*N*	Hypothesis 3		Hypothesis 4	
	35	38	41	44	47	50	53	56	59	62	65	68	71	74	77	80	83	86	89	92	95	98		$\overline{X}_1$ (in μ)	$N\overline{X}_1$	*N*	$N\overline{X}_1$
Unknown	2	2	5	6	6	12	14	9	13	25	9	9	24	6	4	11	5	0	1	.	.	.	163	61.945	.....	..	.....
Hypothesis 3																											
P. contorta 4	.	0.7	9.1	11.9	9.8	16.8	18.2	2.1	1.4	....	....	....	....	....	....	....	...	...	...	.	.	...	70	48.00	336.00	80	384.000
P. ponderosa 7	.	...	....	0.5	0.5	2.5	1.0	2.5	1.5	2.5	8.5	9.0	6.0	5.5	3.0	4.5	0.5	1.0	1.0	.	.	...	50	68.24	341.20	50	341.200
P. radiata 1	.	...	1.0	1.0	1.0	2.0	10.0	10.0	10.0	12.0	8.0	5.0	15.0	3.0	8.0	7.0	6.0	1.0	...	.	.	...	100	65.63	656.30	100	656.300
P. attenuata 30	.	...	....	....	2.0	4.0	10.0	8.0	10.0	30.0	18.0	10.0	4.0	2.0	0	2.0	...	...	...	.	.	...	100	61.64	616.64	100	616.640
P. sabiniana 20	.	...	....	1.0	0.0	0.5	2.0	2.0	0.5	9.0	1.5	8.5	14.5	1.5	4.5	2.5	0.5	0	1.0	0	0	0.5	50	68.33	341.65	71	485.143
Total fitted spectrum	.	0.7	10.1	14.4	13.3	25.8	41.2	24.6	23.4	53.5	36.0	32.5	39.5	12.0	15.5	16.0	7.0	2.0	2.0	0	0	0.5	370	.....	2291.79	401	2483.283
																								$\overline{\overline{X}}_{HO3} = 61.934$		$\overline{\overline{X}}_{HO4} = 61.927$	

P. contorta, P. contorta var. *bolanderi,* and *P. murrayana* are indicated. Actual fitting shows that *P. contorta* 24 is the best.

For *Ab,* with respect to its class-C index of 10-8.5, we have matching reference samples from *P. ponderosa* and from *P. attenuata.* But none of these has a mean as high as 71.00μ (= mean of *Ab,* tables 45, 46). Hence the possibility of this spectrum, *Ab,* being a composite of two components has to be investigated. I therefore formulate from *Ab,* the subspectra *Ab*1 and *Ab*2 (tables 45, 46). Both subspectra

TABLE 45

FITTING OF SPECTRUM *A* OF FOSSIL SAMPLE MERCED 2163

Sample	Pollen frequency by class midpoint																*N*	$\overline{X}$ (in μ)
	38	41	44	47	50	53	56	59	62	65	68	71	74	77	80	83		
Unknown *A* × 2	2	2	4	4	2	18	8	8	6	2	12	18	4	4	2	4	100	61.88
Aa	2	2	4	4	2	18	7	6	4	0	1	..	..	..	..	..	50	52.76
Ab	..	..	..	..	..	..	1	2	2	2	11	18	4	4	2	4	50	71.00
*Ab*1	..	..	..	..	..	..	1	1	1	1	9	1	1	1	1	..	17	68.00
*Ab*2	..	..	..	..	..	..	..	1	1	1	2	17	3	3	1	4	33	72.54

TABLE 46

STATISTICS OF HYPOTHETICAL SPECTRA FORMULATED FROM SPECTRUM *A*, FOSSIL SAMPLE MERCED 2163

Hypothetical spectrum	$\overline{X}$ (in μ)	Mode	Range	Class-C index
Aa	52.76	53	38–68	11–12.24
Ab	71.00	71	56–83	10–8.45
Ab 1	68.00	68	56–80	9–8.54
Ab 2	72.54	71	59–83	9–7.67

*Ab*1 and *Ab*2 have a similar class-C index of 9-8, but the means and modes are different. *Ab*1 has a mean of 68.00μ and unimodal distribution at class 68, and *AB*2 has a mean of 72.54μ and unimodal distribution at class 71. Processing these in the same way as we did for identifying single species (see Part III), I found that samples of *P. sabiniana, P. radiata,* and *P. ponderosa* are eligibles. After a number of trials, *P. sabiniana* was eliminated, but *P. ponderosa* 27 and *P. radiata* 3*A* remain as eligibles.

Now we have some idea of the plausible components of spectrum *A* as a whole. It consists at most of three species. One of these is *P. contorta* 24, which I have used to fit subspectrum *Aa.* The others are *P. ponderosa* 27 and *P. radiata* 3*A,* which may be used to fit the subspectrum *Ab.* It is advisable to find the quantity of each before actually fitting the spectrum. To find the quantity, use equations (1), (2), and (3). Let *A* be the mean of *P. contorta* 24 (= 51.74μ), *B* be the mean of *P. radiata* 3*A* (= 67.88μ), and *C* be the mean of *P. ponderosa* 27 (= 71.48μ).

Let x, y, and z be the quantities of A, B, and C, respectively. The mean of the spectrum is 61.88μ. This is the required mean. For three components, the sum of means to be compared is $3 \times 61.88\mu = 185.64\mu$. Substituting the above values in equations (1) and (3), we have

$$51.74x + 67.88y + 71.48z = 185.64$$

This simplifies into

$$x + 1.312y + 1.3815z = 358.79, \tag{14}$$

and

$$x + y + z = 300.00 \tag{15}$$

Solving (14) and (15) simultaneously, we have

$$0.312y + 0.3815z = 58.79.$$

If $y = 0$, $z = 154.1$; if $z = 0$, $y = 188.4$.

Transform into parametric equations (see p. 64), and let

$$y = 188.4 - t; \text{ then } t = 1.223z, \text{ and } z = 0.8178t.$$
$$x = 300 - 188.4 + t - 0.8178t = 111.6 - 0.1822t.$$

If $t = 0$, $x = 111.6$, $y = 188.4$, $z = 0$.

Since $x = 111.6$ is the minimum limit where $t = 0$, *P. contorta* 24 can be fitted into the spectrum; then z (quantity of *P. ponderosa* 27) $= 0$, and y (quantity of *P. radiata* 3*A*) $= 188.4$ grains. The calculations show that the quantity of *P. contorta* in spectrum *A* cannot be less than 111.6 grains. Even with this minimum amount, the frequency at mode 53 is high. With *P. contorta* taking 111 grains out of a 300-grain sample, the other component must have 190 grains. This component can only be *P. radiata*, because when $t = 0$, $x = 111.6$, $y = 188.4$, and $z = 0$, where z represents the quantity of *P. ponderosa*. Hence the calculations show that we only have two components for spectrum *A*.

Now we pause for a moment to go over what we have found during this second trial to fit the spectrum $A + B$.

We started with spectrum *B* and with little difficulty found that it is made up of five components: *P. contorta*, *P. radiata*, *P. ponderosa*, *P. attenuata*, and *P. sabiniana*. The quantities of each are listed in table 44.

Then we took spectrum *A*. Anticipating that it contains probably two components, we multiplied the frequency in every class by 2 (table 45). Following the same assumption of having two components, spectrum *A* was broken down into two subspectra *Aa* and *Ab*. For *Aa*, *P. contorta* is fitted. For *Ab* we found, however, that both *P. ponderosa* and *P. radiata* were eligible. The subdividing of *Ab* into *Ab*1 and *Ab*2 offered no assistance in fitting. We subjected the data of spectrum *A* as a whole to calculations and found that *P. ponderosa* cannot be the component of subspectrum *Ab*. Hence we conclude that spectrum *A* as a whole is formed by two components only, *P. contorta* and *P radiata*, species we already know to be components of spectrum *B*. Hence spectrum $A + B$ is composed of five components. They are *P. contorta*, *P. radiata*, *P. ponderosa*, *P. attenuata* and *P. sabiniana*.

FITTING OF SPECTRUM $A + B$: THIRD TRIAL

If spectra A and B can be fitted separately, there is no reason why we should not be able to do so when they are combined into one spectrum. The failure we suffered previously was the result of a poor choice to consider *P. jeffreyi* and *P. coulteri.* Now that the components have already been found in our second trial (table 47), and with these five species as guides, we should be able to achieve a full solution. After some trials, we finally come up with a solution in hypotheses 13 and 14 (tables 47, 48).

TABLE 47

COMPARISON OF SAMPLE SIZE OF EACH SPECIES BASED ON ESTIMATES IN THE SECOND AND THIRD TRIALS OF SPECTRUM-FITTING OF MERCED 2163, SPECTRUM $A + B$

Species as components	Hypothesis 4, second trial						Hypotheses in third trial	
	Spectrum B		Spectrum A		Spectrum $A + B$		Spectrum $A + B$	
	Fitted sample size (N_{FB})	True sample size (N_B)[a]	Fitted sample size (N_{FA})	True sample size (N_A)[b]	True sample size		Hypothesis 13	Hypothesis 14
					Total	Percent	(in percent)	
P. contorta.......	80	32	110	18	50	23.5	21.0	23.5
P. ponderosa......	50	20	...	..	20	9.4	13.2	8.8
P. radiata........	100	41	190	32	73	34.3	26.3	29.4
P. attenuata......	100	41	...	..	41	19.2	13.2	8.8
P. sabiniana......	71	29	...	..	29	13.6	26.3	29.4
Total N........	401	163	300	50	213	100.0	100.0	100.0

[a] $N_B = \frac{(N_{FB})(163)}{401}$.

[b] $N_A = \frac{(N_{FA})(50)}{300}$.

In the third trial, we also used the data of overall measurements and obtained a ratio R_4 between corpus and overall size as a supplementary check. Since overall data may be different in numerical composition from those of the corpus, we do not expect a perfect fit of R_4 between the fitted data and that of the unknown. But the difference between them is fairly small, being 0.0346 in hypothesis 13 and 0.0414 in hypothesis 14. Both hypotheses also give good approximations of mean value. That of hypothesis 13 does not differ from the required mean, and that of hypothesis 14 does differ by 0.07μ, though either of them may be considered the answer to the problem.

Comparing the results of the second trial with those of the third (table 47), we note a slight depression of the quantities of certain components. Table 47 has all the quantities under hypotheses from these two trials tabulated in percentages so that they can be compared. The results from fitting A and B separately in the second trial are pooled and compared with those of hypotheses 13 and 14 of the third trial. The picture of *P. contorta* and *P. ponderosa* seems to be quite in ac-

TABLE 48

FITTING OF SPECTRUM $A + B$, FOSSIL SAMPLE MERCED 2163: HYPOTHESES 13 AND 14

Sample	Pollen frequency by class midpint																										
	35	38	41	44	47	50	53	56	59	62	65	68	71	74	77	80	83	86	89	92	95	98	N	$\overline{X}_1$ (in μ)	$(N\overline{X}_1)$	$\overline{X}_3$	$(N\overline{X}_3)$	R_4
Unknown	2	3	6	8	8	13	23	13	17	28	10	15	33	8	6	12	7	0	1	..	..	..	213	61.93	......	86.560	......	1.3977
Hypothesis 13																												
P. attenuata 11	.	...	...	...	....	1.5	3.0	3.5	5.0	16.0	15.5	3.5	1.0	....	....	....	...	...	...	...	..	..	50	61.91	309.55	87.83	439.15	......
P. contorta 23	.	1.6	4.8	6.4	12.8	19.2	25.6	4.8	3.2	1.6	....	....	....	....	....	....	...	...	...	...	...	...	80	50.18	401.44	64.34	514.72	......
P. ponderosa 31	.	...	...	...	1.0	1.0	3.0	6.0	7.0	22.0	7.0	3.0	....	....	....	....	...	...	...	...	...	...	50	60.56	302.80	90.08	450.40	......
P. sabiniana 2D	.	...	...	2.0	.0	1.0	4.0	4.0	1.0	18.0	3.0	17.0	29.0	3.0	9.0	5.0	1.0	.0	2.0	.0	.0	1.0	100	68.33	683.30	94.94	949.40	......
P. radiata 1	.	...	1.0	1.0	1.0	2.0	10.0	10.0	10.0	12.0	8.0	5.0	15.0	3.0	8.0	7.0	6.0	1.0	...	...	...	...	100	65.63	656.30	85.42	854.20	......
Total	.	1.6	5.8	9.4	14.8	24.7	45.6	28.3	26.2	69.6	33.5	28.5	45.0	6.0	7.0	12.0	7.0	1.0	2.0	0	0	1.0	380	61.93	2353.39	84.12	3207.87	1.3631
Hypothesis 14																												
P. attenuata 11	.	...	...	...	....	0.9	1.8	2.1	3.0	9.6	9.3	2.1	0.6	....	....	....	...	...	...	...	...	...	30	61.91	185.73	87.83	270.49	......
P. contorta 23	.	1.6	4.8	6.4	12.8	19.2	25.6	4.8	3.2	1.6	....	....	....	....	....	....	...	...	...	...	...	...	80	50.18	401.44	64.34	514.72	......
P. ponderosa 31	.	...	...	...	0.6	0.6	1.8	3.6	4.2	13.2	4.2	1.8	....	....	....	....	...	...	...	...	...	...	30	60.58	181.68	90.08	270.24	......
P. sabiniana 2D	.	...	...	2.0	0	1.0	4.0	4.0	1.0	18.0	3.0	17.0	29.0	3.0	9.0	5.0	1.0	0	2.0	0	0	1.0	100	68.33	683.30	94.94	949.40	......
P. radiata 1	.	...	1.0	1.0	1.0	2.0	10.0	10.0	10.0	12.0	8.0	5.0	15.0	3.0	8.0	7.0	6.0	1.0	...	...	...	...	100	65.63	656.30	85.42	854.20	......
Total	.	1.6	5.8	9.4	14.4	23.7	43.2	24.5	21.4	54.4	24.5	25.9	44.6	6.0	17.0	12.0	7.0	1.0	2.0	0	0	1.0	340	62.00	2108.45	84.09	2859.05	1.3563

cord in all hypotheses. The overall comparison seems to indicate that hypothesis 14 is in better accord with results of the second trial than hypothesis 13. Hence hypothesis 14 is chosen as the solution.

Note that all attempts led essentially to the same components.

FITTING OF SPECTRUM *C*

Spectrum *C* (table 38) has an even spread, a long span, a mode at class 53, and possibly the same minor modes if we had a larger sample size. The class-C index is very high (17-22). Lot *C* is conceivably made up of three or more components or of two species that differ widely in their means and ranges. Yet the sample size is only 50, not enough in this case for iteration. The low mean of 51.56μ precludes many species, for among the eligibles we have (table 41), only *P. contorta* 23, and no other, has a mean value lower than the sample. Spectrum *C*, then, most probably consists of many *P. contorta* grains and a few grains of some other species. The low mean of 51.56μ is a difficult one to match with only two components, for if one species has a mean, for instance, of 60.00μ, the other will have to have a mean as low as 42.00μ, so that equal sample size of both will yield an average mean of 51.00μ. Samples with low mean are few in the references, and *P. contorta* var. *latifolia* 23 is the only one that has a mean (50.18μ) lower than the required mean of lot *C* (51.56μ) (table 41). If *P contorta* var. *latifolia* 23 should be one of the two components, then the other component must have a mean within the confidence interval of 53.00μ. Even this is difficult to find. From table 41, we find that there is only one sample of only one other species which fits this requirement: *P. attenuata* 24. Unfortunately all possibilities from samples of *P. attenuata*, including 24, become impossibilities when we find that the range of their spectra reaches only class 71, three classes short of the spectrum *C*.

In order to account for a spectrum span reaching class 80, we need a sample having a higher mean. Let this first component have a mean of 60.00μ. If the second component has an equal share in frequency, it has to have a mean of 42μ. But if the second component has more grains, and let this second component be *P. contorta* var. *latifolia* 23, then, in a total sample size of 100, the first component cannot have more than 15 grains. It would be fine if we could find a reference sample with suitable mean, range, and frequency distribution to go with *P. contorta* var. *latifolia* 23 and fit nicely the average mean of spectrum *C*. After considerable effort, the best estimate I can possibly achieve is a composite spectrum of three components. These are *P. contorta,* 74 grains, *P. attenuata,* 24 grains, and *P. radiata,* 2 grains. The spectrum gives an average mean of 52.12μ, not a very good fit of the required mean of 51.56μ. Nor have I succeeded in fitting the frequency of class 32.

CONCLUSIONS

This Merced sample worked out in detail is a fossil sample and we have no way of proving whether or not the identification of its components is true to nature. The process of identification is essentially one of deduction. When different analytical approaches invariably lead to the same taxa, our confidence in the process and its results is thereby strengthened.

The good result achieved in approximating the mean should be pointed out. In

TABLE 49

FITTING BY SIMULATION OF SPECTRUM OF FOSSIL SAMPLE MERCED 3022

Pollen sample	N	$\overline{X}_1$ (in μ)	Pollen frequency by class midpoint																		Class number	C (in percent)
			41	44	47	50	53	56	59	62	65	68	71	74	77	80	83	86	89	92		
Fossil: Merced 3022	100	71.03	2	0	0	5	6	0	8	8	3	7	17	5	8	9	6	9	5	2	18	17.21
Hypothetical component																						
A1 *P. lambertiana*	14	51.71	2	0	0	5	4	0	2	1	..	..	..	..	..	..	..	..	..	..	8	11.5
*B*1	25	66.68	..	..	..	..	2	0	5	5	0	1	7	1	2	1	1	..	..	..	11	12.2
*C*1	61	77.25	..	..	..	..	..	..	1	2	3	6	10	4	6	8	5	9	5	2	12	10.8
*B*2	28	68.75	..	..	..	..	2	0	5	5	0	1	7	1	2	1	1	3	..	..	12	14.3
*C*2	58	76.79	..	..	..	..	..	..	1	2	3	6	10	4	6	8	5	6	5	2	12	10.9
*B*3	36	69.42	..	..	..	..	2	0	5	6	3	3	7	1	2	1	1	3	1	1	14	14.7
*C*3	50	77.60	..	..	..	..	..	..	1	1	0	4	10	4	6	8	5	6	4	1	12	9.4
*C*4 *P. jeffreyi*	51	76.12	..	..	..	..	..	..	2	2	0	6	11	5	3	8	4	5	4	1	12	10.8
*B*4 unknown	35	71.34	..	..	..	..	2	0	4	5	3	1	6	0	5	1	2	4	1	1	14	14.1
*B*4 × 3	105	.....	..	..	..	..	6	0	12	15	9	3	18	0	15	3	6	12	3	3		
B4a	43	.....	..	..	..	..	..	..	3	2	3	1	1	0	12	3	6	7	3	2	*P. sabiniana*	
B4b	30	.....	..	..	..	..	6	0	0	9	1	0	10	0	3	0	0	1	..	..	*P. radiata*	
B4c	16	.....	..	..	..	..	..	..	4	2	3	1	4	0	0	0	2	..	..	..	*P. ponderosa*	
B4d	16	.....	..	..	..	..	..	..	5	2	2	1	3	0	0	0	0	2	0	1	*P. attenuata*	

frequency distribution, some weakness still needs to be overcome. For instance, in hypothesis 14 (table 48), the mode at class 71 is, proportionally speaking, still lower than that required. Mode at class 80 cannot be matched unless *P. jeffreyi* or *P. coulteri* is involved. We must realize that iteration may account for the major ingredients, but not necessarily for the minor ones. The estimated number of components probably will always be short of the real number in fossil assemblages. The ones missed will be those having fewer pollen grains in the assemblage to begin with. In short, iteration, as I have used it, will give most, but not all, of the picture of fossil content.

Example II: Fossil Sample 2, Merced 3022

The sample from Merced locality 3022 contains pine pollen that yield the data shown in table 49 (top row). It was studied and identified by a procedure different from the one used with Merced 2163.

The spectrum for Merced 3022 has 18 classes, and $C = 17.21$ per cent. From the reference data, no single species has equivalent statistics; hence the sample must be a mixed one.

Morphology shows that there are a number of grains with verrucose distal membrane. These are tentatively identified as *P. lambertiana.*

Other grains look rather like *P. sabiniana* and *P. jeffreyi* in the sacci attachment of most of their grains. Their corpi are also large. The high sample mean of 71.03μ is caused by the prevalence of these large-sized grains.

Initial grouping and fitting *A* and *C* spectra

The Merced 3022 sample was split into three groups of different sizes. Their spectra were labeled *A, B,* and *C,* grouped in ascending order of mean values, and each of the three was labeled 1, 2, 3, and up, indicating successive approximations (table 49). The spectra overlap one another, for it is necessary to maintain a minimum span to simulate a reasonable spectrum. For instance, reference samples show that a spectrum beginning at class 50, and having a mean compatible with that of the fossil spectrum *B,* usually has 11 to 12 classes. Spectrum *B* is therefore given 11 classes. The span of spectrum *C* is similarly determined.

Spectrum *A*1 consists solely of grains with sculptured distal membrane. It totals only 14 grains. *P. lambertiana* 4 seems to be the best fit for mean, range, and modal distribution. The large sacci and Y pattern of the cappa also indicate *P. lambertiana.*

Spectrum *C*1 at the higher end of the spectrum consists of larger grains. The high mean value, 77.25μ, can be matched only by reference samples from three species: *P. jeffreyi, P. coulteri,* and *P. sabiniana.* The spectrum *C*1 has its major mode at class 71 and a possible minor mode at class 86. While mode at 71 is found among samples of *P. jeffreyi* and *P. coulteri,* modes at 80 and 86 are found only in samples 4 and 11 of the latter species. The range of *C*1 is very much shorter than *P. coulteri* 11, and there is no mode at 71 on the spectrum of *P. coulteri* 4. A part of the high frequency that occurs at class 86 must therefore be interpreted as from some other species. But there is no other species among the references to match. Hence the spectra *C*1 and *B*1 are readjusted by deducting three grains from the $C1$ spectrum and adding them to the $B1$ spectrum. The spectra of $B1$ and $C1$ thus changed appear as $B2$ and $C2$ in table 50.

The mean and C are then recalculated for $B2$ and $C2$. The class-C index for spectrum $C2$ is normal for a large number of species and samples, but the high mean value of 76.79μ indicates only *P. jeffreyi*. Among the samples of *P. jeffreyi,* 1, 12, 15, 16, and 28 are all eligible, but only 12 has an identical modal distribution.

Spectrum $B2$ is bimodal. The modes are at classes 71, 62-59. Since these two modes never occur together in the references, unless the spectrum is trimodal and accompanied by a third mode at class 77 or 80 (see table 40, row 4), it is not

TABLE 50

TWO FITTINGS OF SPECTRUM $B4$, N_1 AND N_2, FOSSIL SAMPLE MERCED 3022

Species and sample number	First fitting		$\overline{X}_1$ (in μ)	Second fitting		
	N_1	$N_1\overline{X}_1$ (in μ)		N_2	$N_2\overline{X}_1$ (in μ)	Percent
P. radiata 1	33.0	2165.79	65.63	30	1968.90	10.00
P. ponderosa 25	15.5	1054.00	68.00	16	1088.00	5.33
P. attenuata 4	15.5	1051.68	67.85	16	1085.60	5.33
P. sabiniana 4	41.0	3192.67	77.87	43	3348.41	14.34
Total	105.0	7464.14	.....	105	7490.91	35.00
Required $B5$	105.0	7490.82	.....	105	7490.82	35.00
Difference	.....	−25.68	.....	...	−0.09	.....

Adjusted: + 100 percent of *P. sabiniana* 4 = 77.87
− 100 percent of *P. radiata* 1 = 65.63
Net gain for exchange of 100 percent = 12.24
Net gain for exchange of 1 per cent = 0.1224

possible to match the spectrum of $B2$ with any reference. *P. attenuata* 25, *P. radiata* 1 and 3*A*, and *P. sabiniana* 20 have modal distributions that most closely resemble those of the assumed spectrum $B2$. The class-C index of 12-14, however, is not matchable by any references. Successive adjustments of the spectra B and C, through $B2$, $C2$, $B3$, $C3$ to their final form at $B4$ and $C4$, are therefore made. This process of readjustment goes on until one of the assumed spectra yields a normal class-C index. The class-C index of $C4$ is 12-11, a normal one, indicating also *P. jeffreyi.* (The range of $C4$ is from class 59 to class 92. The eligibles with compatible mean, 76.12μ, are *P. jeffreyi* and *P. coulteri*. Under the class-C index, *P. jeffreyi* and *P. sabiniana* are indicated, with *P. jeffreyi* being the only one common to both lists.)

Up to now spectrum A is believed to be pollen grains of *P. lambertiana,* and spectrum C, to be pollen grains of *P. jeffreyi.*

COMPONENTS OF SPECTRUM $B4$

Spectrum $B4$ is still a heterogeneous one. The class-C index is 14-14 and has no match in the references. It consists of 35 grains. It would be more convenient to work on it if we had more grains. I transformed the spectrum into one that contains 105 grains by multiplying the original 35 grains by three. From this spectrum a number of subspectra are again formed—*B4a, B4b, B4c,* and *B4d* (table 49)—using essentially the same process of successive simulation as was used for breaking down spectra *A*, *B*, and *C*.

Spectrum *B4* has four modes: 71, 62-59, 77, 86. From known distribution of modes, a 71, 62-59, 77 combination is the typical trimodal distribution of *P. radiata;* a 71, 62, 77, and 77-71 combination matches that of *P. sabiniana.* Deducting the possible number of grains of these two species from spectrum *B4* and forming *B4a, B4b,* there remains a difference of 32 grains having modes at 65 and 71. The mode at 65 is typical of many samples of *P. ponderosa* and *P. attenuata.*

Frequencies at 65, 71 belong to *P. attenuata* 4, and those at 65, 80, 86 are equivalent to a number of *P. ponderosa* samples.

Table 50 shows the data of subspectra of *B4* and their fittings. The required sample size mean product ($N\overline{X}$) of *B4* is 7490.82μ; the $N_2\overline{X}$ of the estimated fitting is 7490.91μ, a good fit. $N_1\overline{X}$ is also shown, but is not as good.

The reason the species and samples were chosen for fitting is based not only on modal distribution, but also on the estimated means of spectrum *B4*, by trial-fitting the sample mean of 71.03μ with chosen references. From this, the quantities of different species are adjust to achieve the fit.

For *B4* the results are: 10.0 per cent of *P. radiata,* 5.3 per cent of *P. ponderosa,* 5.3 per cent of *P. attenuata,* and 14.4 per cent of *P. sabiniana* (table 52), a total of 35 per cent.

A word to explain the calculations involved in obtaining these results.

Two fittings of spectrum *B4*, N_1 and N_2, are shown in table 50. The number of grains fitted into each species made a total of 105 grains. This total should be equivalent to 35 per cent of the entire sample size of the fossil sample Merced 3022, that is, spectrum $A + B + C = 100$ per cent. We know from our fitting that *A1* is 14 per cent and *C4* is 51 per cent; hence *B4* must be 35 per cent. We also know the means of *A1* and *C4*. We want to find the required mean of spectrum *B4* so that the sum of means of *A, B,* and *C* will average 71.03μ, the mean of the total sample Merced 3022. In table 51, the calculation for this purpose is shown. Multiply the mean ($\overline{X}$) by the respective sample size (N). The product is $N\overline{X}$. Multiply the mean of the total sample by 100, resulting in a size-mean product of 7103.00μ. Then subtract from this the size-mean product of *A1* and *C4*. The difference is then the required size-mean product of *B4*, shown in table 51, note, as $N_w\overline{X}_1$, and it equals 7490.82μ.

In table 50 are entered the size-mean products of each of the species fitted in spectrum *B4*, on the basis that the sample size of *B4* equals 105. Fitting N_1 differs slightly from fitting N_2 in the relative quantity of the four species. The size-mean products are summed (Σ), giving $\Sigma N\overline{X}$ for each fitting. The summation of size-mean products of N_1 and N_2 is then compared with the required value ($N_w\overline{X}$ in table 51). That of N_1 differs by 25.68μ, that of N_2 differs by only 0.09μ. Hence we accept the fitting of N_2 as correct. Transform the N_2 value into 35 per cent by making the total value of 105 equal to 35 per cent, and obtain the quantity of each species as percentages (table 50, last column).

The slight difference of 0.09μ is negligible (see calculation, bottom of table 52). To change one grain from one species to another species changes the size-mean product by 0.1224μ. The difference one grain makes is more than the difference found by the N_2 fitting, which is only three-fourths of a grain. Hence, we retain the fitting N_2 as final.

TABLE 51

PROCEDURES IN SPECTRUM-FITTING FOSSIL SAMPLE MERCED 3022

Spectrum	Species or component	N	$\overline{X}_1$ (in μ)	$N\overline{X}_1$ (in μ)
*A*1	*P. lambertiana*	14	51.71	723.94
*C*4	*P. jeffreyi*	51	76.12	3883.12
*A*1 + *C*4	....	65	.....	4606.06
*B*4	Unknown	35	.....	2496.94
Total *A* + *B* + *C*		100		7103.00

NOTE: Multiply the required N of B_4 by 3 = 105; call this N_w. Then $3N\overline{X}_1$ = (2496.94)(3) = 7490.82 = $N_w\overline{X}_1$.

TABLE 52

SAMPLE SIZE AND PERCENTAGE OF VARIOUS COMPONENTS FITTED FOR FOSSIL SAMPLE MERCED 3022

Spectrum	Species fitted	N	Percent
*A*1	*P. lambertiana*	91	14
*C*4	*P. jeffreyi*	331	51
*B*4	*P. radiata*	65	10
*B*4	*P. ponderosa*	35	5.3
*B*4	*P. attenuata*	35	5.3
*B*4	*P. sabianiana*	93	14.4
Total *A* + *B* + *C*		650	100.0

COMPONENTS AND THEIR QUANTITIES

A total of 650 grains of pine pollen was counted in fossil sample Merced 3022. They are now known to be divided in unequal proportions among six species. Calculations show there are, in round numbers:

P. jeffreyi, 331 grains	*P. ponderosa,* 35 grains
P. lambertiana, 91 grains	*P. attenuata,* 35 grains
P. radiata, 65 grains	*P. sabiniana,* 93 grains

The mean of the unknown spectrum is 71.03μ. The mean of these components is 71.03μ (table 52), and all the modal distributions fit.

OTHER CONSIDERATIONS

If we use spectrum *B*3 instead of *B*4 we find that *P. radiata* and *P. attenuata* still are components. Subtracting these grains from the spectrum yields a difference that makes no sense. In that event we have to include all the remainders in spectrum *C* of *P. jeffreyi.* By including these grains in spectrum *C,* it is then changed into a mixed spectrum indicated by the number of classes and *C.* By breaking down the newly made spectrum *C,* it will again resolve into *P. jeffreyi, P. sabiniana,* and *P. ponderosa,* and we obtain the same conclusions arrived at from our study of *B*4.

SUMMARY

In the attempt to identify the species of pines from their pollen grains, statistics of some 280 samples of pollen grains of pines were studied and analyzed. The morphological features of the pine pollen grains supplied two criteria by which species of pine may be subdivided into groups, recognized from the character of their distal membranes: Group A has a verrucoid distal membrane, whereas Group B has a psilate distal membrane. Other subdivisions may be made from the two patterns of the cappa, the Y pattern and the W pattern. Since the usefulness of morphology is limited, statistics are depended upon mainly for specific classification.

Among the statistics used, emphasis is laid on the ratios between the mean measurements of the breadths of corpus, of saccus, and of overall dimension of the pollen grain. The ratios are converted into a cos A index, which in the working examples appears to be the discriminant of species.

In identifying fossil pollen grains, the combined index of number of classes in a spectrum and the coefficient of variance are employed as criteria for heterogeneity. The modal distribution of the frequencies formed the major basis for specific identification through iteration.

The method has been tested, and 90 per cent of modern unknowns supplied by others have been correctly identified by this method. There is yet no way to test its application to fossil grains. Since iteration is an elementary method, much of course, remains to be done on this part of the problem.

The technique used is acetolysis for modern reference material and hydrofluoric acid treatment and acetolysis for fossil material. The sizes of pollen grains so treated are comparable.

Programing by discriminant analysis shows that statistics are useful in classifying the pollen grains into their respective species. A program of weighted parameters and gamma indexes also reached a similar conclusion.

APPENDIX I

Statistics of Individual Reference Samples of Modern *Pinus* Pollen Grains

Appendix table 1 contains statistical data of modern pollen grains of species of *Pinus*. The data included are mean of corpus breadth ($\overline{X}_1$), mean of saccus breadth ($\overline{X}_2$), mean of overall size ($\overline{X}_3$), all measured in μ; the class number, the coefficient of variance, *C*, in per cent; the range of sizes, the standard deviation, *SD* in μ; the ratio index, and the cos A index. The class, *C*, range, and *SD* all are derived from measurements of the corpus only. The ratio index is derived from $\overline{X}_1/\overline{X}_2$ and $\overline{X}_3/\overline{X}_1$, and the cos A index is derived from all three means ($\overline{X}_1$, $\overline{X}_2$, and $\overline{X}_3$). Range is usually expressed by the upper or maximum end and the lower or minimum end of the spectrum. Since I use grouped data, I identify these endings by the respective class midpoints. The figures listed in the Range columns are codes. The code base is at 0, where the midpoint is 25μ. A code number can be transformed into midpoint readings by adding the number six times the code number to the base 25μ.

There are eighteen species, four varieties, and three hybrids of *Pinus* in the list, all arranged alphabetically by species. A few doubtful samples are indicated by an asterisk after the sample number.

APPENDIX TABLE 1

STATISTICS OF INDIVIDUAL REFERENCE SAMPLES OF MODERN *Pinus* POLLEN GRAINS
(In alphabetical order by species)

Species and sample number	X_1 (in μ)	X_2 (in μ)	X_3 (in μ)	Class number	C (in percent)	Range Min. (in code)	Range Max. (in code)	SD (in μ)	Ratio index	Cos A
P. albicaulis										
1	55.58	60.05	82.57	12	16	2	8	8.81	914	0.7396
2	55.70	60.12	88.62	8	14	4	7	8.15	815	0.7851
7	59.54	56.12	88.76	15	14	2	9	8.27	1014	0.7511
9	60.68	55.28	91.76	9	10	4	8	8.34	1015	0.7682
10	55.64	56.96	86.84	7	8	4	7	4.56	913	0.7773
11	60.11	59.00	90.86	12	11	3	8	6.56	1015	0.7577
13	64.31	57.89	90.38	10	8	4	9	5.43	1114	0.7056
14b	60.92	55.52	91.70	9	10	4	8	6.30	1015	0.7641
B7	53.60	53.13	80.90	10	11	2	7	6.12	1015	0.7564
P. aristata										
1	45.05	40.70	68.18	9	12	1	5	5.54	1114	0.7704
2	48.11	45.02	66.53	9	10	2	6	5.06	1013	0.6909
3	52.91	49.88	76.61	11	10	2	7	5.58	1014	0.7272
6A	54.38	53.30	81.40	14	12	1	8	6.68	1014	0.7502
6B	53.69	53.30	81.40	11	12	2	7	6.71	1014	0.7586
7	53.00	47.96	74.30	9	10	3	7	5.25	1114	0.7032
8	50.24	44.54	70.98	11	11	2	7	5.68	1114	0.7114
9	48.20	49.82	71.54	9	11	2	6	5.52	914	0.7403
10	47.06	50.90	66.32	8	10	2	5	4.86	914	0.7072
14	54.08	49.58	71.54	8	8	3	7	4.32	915	0.6557
15	50.18	52.01	77.00	9	10	2	6	5.20	1013	0.7636
16	54.52	51.74	68.30	10	8	3	7	4.60	1013	0.6182
17	49.58	46.07	66.53	8	8	2	6	4.02	1013	0.6673
P. attenuata										
1	58.88	51.14	80.32	11	10	3	8	6.22	1113	0.6816
2	59.24	51.41	88.10	11	10	4	9	5.86	1114	0.7612
3	57.41	48.53	81.18	10	9	3	8	5.26	1114	0.7170
4AD	67.85	56.78	92.81	15	8	4	11	5.59	1113	0.6864
7	64.07	56.77	91.66	16	13	4	11	8.53	1114	0.7225
10	59.93	49.61	87.68	12	10	4	8	5.81	1214	0.7537
11	61.91	50.30	87.83	8	7	4	8	4.44	1214	0.7256
12	64.46	54.20	93.70	10	7	4	9	4.70	1114	0.7445
13	64.19	52.91	87.14	14	11	3	10	6.88	1213	0.6802
14	59.48	53.42	87.62	11	9	3	8	5.47	1114	0.7470
16	59.57	51.98	86.01	13	12	3	9	7.13	1114	0.7327
17	62.42	52.31	86.61	11	9	4	9	5.89	1113	0.6998
18	63.74	51.96	85.30	10	7	5	9	4.65	1213	0.6671
19	63.89	53.18	86.84	9	9	5	9	5.62	1213	0.6807
20	58.52	48.08	85.10	10	9	3	8	5.29	1214	0.7490
21	57.29	49.97	77.42	11	10	3	8	5.71	1113	0.6732
22	63.23	55.61	88.76	12	9	4	9	6.31	1114	0.7063

APPENDIX TABLE 1—*Continued*

Species and sample number	X_1	X_2	X_3	Class number	C (in per-cent)	Range		SD (in μ)	Ratio index	Cos A
	(in μ)					Min. (in code)	Max.			
23	60.68	52.00	84.23	11	10	3	8	6.23	1113	0.6968
24	56.96	52.01	80.99	12	10	2	8	5.92	1014	0.7146
25*	69.14	60.44	102.14	9	9	5	9	5.94	1114	0.7539
26*	69.76	56.96	95.66	9	9	5	9	5.98	1213	0.6909
27	67.70	61.88	100.28	10	8	5	10	5.56	1014	0.7495
28	62.78	59.72	96.08	10	9	4	9	5.61	1015	0.7718
29N*	69.00	59.48	101.30	14	12	4	11	8.11	1114	0.7501
30	61.64	56.12	91.22	12	10	4	9	6.38	1213	0.7492
30T	59.45	52.60	86.26	8	8	4	8	4.68	1014	0.7348
31*	69.47	67.52	104.72	14	8	4	10	5.83	1015	0.7565
32	58.64	52.76	96.92	9	9	3	7	5.26	1116	0.8544
33	64.10	58.04	90.74	8	9	5	8	5.50	1114	0.7114
34*	70.59	66.74	102.80	11	9	5	10	6.10	1014	0.7316
36	66.08	59.60	96.14	10	8	5	9	5.51	1114	0.7355
Pinus attenuata X radiata										
9	64.94	54.08	89.36	8	7	5	8	4.57	1213	0.6925
P. balfouriana										
2	50.28	42.80	66.87	11	10	2	7	5.14	1113	0.6596
4A	50.96	43.42	66.91	10	12	2	7	6.08	1113	0.6480
4H	51.32	45.74	65.48	10	11	2	7	5.48	1112	0.6254
5	58.70	48.86	71.28	10	12	3	7	7.30	1212	0.5775
6	51.44	44.06	65.48	9	9	2	6	4.72	1112	0.6209
8	55.52	50.06	73.94	9	9	3	7	5.15	1113	0.6606
9	53.72	51.80	75.02	11	10	2	7	5.64	1013	0.6981
18	54.23	49.28	73.16	10	10	2	7	5.80	1113	0.6712
P. contorta										
1	52.43	40.40	70.61	5	6	3	6	3.21	1213	0.6781
2	58.52	45.98	77.81	8	7	4	7	3.80	1213	0.6630
5	61.91	54.42	86.98	11	9	4	9	5.76	1114	0.7071
6	52.36	43.07	71.00	7	7	3	6	3.74	1213	0.6793
7	58.71	50.18	80.45	13	9	3	9	5.56	1113	0.6865
11	52.82	47.72	74.90	9	6	3	7	3.40	1114	0.7131
12	51.14	44.60	73.60	10	11	2	7	5.67	1114	0.7297
14	56.42	40.20	74.00	10	9	3	8	4.47	1413	0.7250
15	48.44	38.90	67.10	8	10	3	6	4.78	1213	0.7029
16	49.04	38.72	65.42	9	12	2	6	5.67	1213	0.6660
17	56.24	48.86	78.44	8	8	4	7	4.77	1113	0.7015
18	56.72	44.90	78.20	6	7	4	7	3.95	1213	0.6998
19	51.91	46.80	75.58	10	11	3	7	5.64	1114	0.7362
20	51.42	45.28	73.52	6	7	3	5	3.42	1114	0.7226
22A	56.42	49.82	83.48	9	10	3	7	5.39	1114	0.7535
22B	57.74	49.82	83.48	7	9	4	7	5.21	1114	0.7354
24	51.74	45.32	74.36	9	9	3	6	4.58	1114	0.7279

APPENDIX TABLE 1—*Continued*

Species and sample number	X_1 (in μ)	X_2 (in μ)	X_3 (in μ)	Class number	C (in per-cent)	Range Min. (in code)	Range Max. (in code)	SD (in μ)	Ratio index	Cos A
25	54.68	41.84	72.86	9	10	4	7	5.22	1313	0.6674
26	44.96	35.96	60.80	6	9	3	5	4.22	1213	0.6788
27	51.80	39.14	67.34	7	9	2	5	4.89	1313	0.6418
22	57.11	49.82	83.48	9	9	3	7	5.30	1114	0.7441
P. contorta var. *bolanderi*										
3	51.89	44.57	65.96	10	10	2	7	4.98	1112	0.6199
4	48.12	39.02	65.24	12	12	1	7	5.88	1213	0.6802
5	52.28	40.58	67.58	7	7	3	6	3.46	1212	0.6346
6	52.46	41.72	69.44	8	8	3	6	4.04	1213	0.6576
7	49.88	41.54	65.96	7	8	3	6	4.15	1213	0.6548
P. contorta var. *latifolia*										
8	53.69	38.39	74.88	7	8	3	6	4.20	1313	0.7302
23	50.18	36.86	64.34	9	10	2	6	4.86	1312	0.6283
25	48.14	39.14	64.94	7	10	2	5	4.82	1213	0.6751
26	52.82	44.42	74.24	9	11	3	7	5.79	1114	0.7118
P. coulteri										
1	72.74	64.97	105.65	16	20	4	12	14.11	1114	0.7351
3NT	86.10	73.68	129.08	13	10	7	13	8.81	1114	0.7716
4	78.88	70.70	109.22	11	11	7	12	8.44	1113	0.6932
5	74.96	65.88	107.11	8	9	4	11	7.07	1114	0.7225
6	75.68	62.78	103.70	11	10	6	11	7.64	1213	0.6887
8	65.72	58.76	93.92	15	14	4	11	8.90	1114	0.7207
9	84.19	75.38	124.36	8	9	8	12	7.64	1114	0.7502
10	76.36	61.80	101.45	13	12	5	11	9.23	1213	0.6604
11	76.89	63.20	101.42	13	13	5	11	9.97	1212	0.6528
12T	73.52	59.90	103.04	9	6	6	10	4.59	1214	0.7129
P. flexilis										
2	55.25	47.12	73.67	11	11	2	7	6.32	1113	0.6619
5	57.80	57.44	84.20	11	11	3	8	6.38	1014	0.7286
6	55.52	53.96	83.60	11	12	3	8	6.79	1015	0.7557
7	58.04	54.92	81.88	12	13	3	8	7.36	1014	0.7063
8T	53.09	50.48	78.20	11	12	2	7	6.64	1014	0.7403
2N	52.01	47.18	74.88	10	10	3	7	5.41	1014	0.7258
9W (Wilson)	68.48	67.82	105.29	12	9	4	10	6.13	1015	0.7732
P. jeffreyi										
2	75.08	67.25	108.07	19	12	5	11	9.04	1014	0.7268
5	87.95	67.77	124.54	20	12	7	13	10.63	1214	0.7237
6	59.08	46.01	88.26	16	13	2	9	7.82	1214	0.7900
7	68.84	58.34	98.00	17	14	3	11	9.78	1114	0.7231
8	72.95	62.15	100.91	12	9	5	11	6.62	1113	0.6955

APPENDIX TABLE 1—*Continued*

Species and sample number	X_1 (in μ)	X_2 (in μ)	X_3 (in μ)	Class number	C (in per-cent)	Range Min. (in code)	Range Max. (in code)	SD (in μ)	Ratio index	Cos A
9T	77.78	67.58	108.92	14	11	5	11	8.31	1114	0.7051
10r	86.15	73.09	117.88	16	10	6	13	8.32	1113	0.6857
11	61.22	48.92	86.42	9	12	4	8	7.34	1213	0.7421
11M	63.65	52.26	88.10	17	13	4	9	8.24	1213	0.6997
12N	74.14	68.30	109.87	16	12	4	12	8.89	1014	0.7489
12	76.10	68.78	109.76	12	9	6	11	6.82	1114	0.7277
14	72.92	65.18	103.64	19	17	4	11	12.10	1114	0.7159
15	73.76	63.98	99.92	13	10	5	11	7.75	1113	0.6755
26	80.06	70.88	118.94	13	9	6	12	7.33	1114	0.7568
26V	75.95	66.35	110.70	12	9	5	12	7.16	1114	0.7412
27	77.18	72.92	114.62	9	7	6	11	5.23	1014	0.7477
28	72.56	61.73	107.12	13	9	5	10	6.56	1114	0.7577
30	72.68	64.44	102.62	15	12	5	11	8.51	1114	0.7106
31	73.34	61.32	100.51	9	9	5	9	6.30	1213	0.6882
P. jeffreyi X *coulteri*										
16	73.09	59.99	97.79	13	10	6	0	7.18	1213	0.6664
P. jeffreyi X *ponderosa*										
20	68.27	63.05	96.20	13	11	4	10	7.48	1014	0.7064
21	71.45	62.15	105.11	13	8	4	10	5.96	1114	0.7505
22	76.25	63.02	102.56	15	10	4	11	7.64	1213	0.6712
22B	76.49	62.24	102.56	12	8	6	12	5.84	1213	0.6691
P. lambertiana										
1	59.00	63.08	87.83	13	15	3	9	8.18	914	0.7411
2	71.06	68.53	100.55	15	12	4	11	8.43	914	0.7081
3	57.98	62.82	88.62	10	10	4	8	5.78	915	0.7579
4	51.75	52.95	79.74	9	9	2	6	4.69	915	0.7679
5	70.52	71.30	107.72	13	12	5	11	8.37	915	0.7626
6	65.63	70.40	98.33	11	10	4	9	6.62	915	0.7452
7	72.70	70.30	99.20	14	12	4	9	8.80	1013	0.6809
8	70.85	70.25	106.16	13	11	4	10	8.05	1015	0.7499
9A	62.21	62.54	90.98	12	13	4	9	8.23	914	0.7310
P. monophylla										
5	49.58	38.18	66.00	9	10	2	6	5.00	1213	0.6658
6	55.64	45.62	75.74	10	11	3	7	6.40	1213	0.6833
7	55.22	46.94	78.26	9	9	4	8	5.06	1114	0.7185
11	59.54	47.06	81.41	10	9	4	8	5.35	1213	0.6913
12	51.32	42.32	71.54	9	8	2	6	4.48	1213	0.7060
13	53.90	41.44	73.18	10	11	2	7	6.25	1213	0.6871
14T	61.13	47.12	82.94	12	8	3	8	5.14	1213	0.6861

APPENDIX TABLE 1—*Continued*

Species and sample number	X_1 (in μ)	X_2 (in μ)	X_3 (in μ)	Class number	C (in per-cent)	Range Min. (in code)	Range Max. (in code)	SD (in μ)	Ratio index	Cos A
P. monticola										
2	56.06	57.47	77.87	12	14	3	8	7.30	913	0.6954
4	53.26	46.13	72.32	9	11	3	7	5.71	1113	0.6776
5	57.41	54.55	84.90	9	8	3	8	4.56	1014	0.7436
6	61.40	56.66	90.68	9	8	4	8	4.89	1014	0.7458
7	63.68	58.04	85.40	11	10	4	9	6.89	1013	0.6664
8	57.56	56.42	79.14	9	10	4	8	6.07	1013	0.6868
9	57.92	55.64	83.24	10	10	2	8	6.26	1014	0.7201
11	58.52	54.86	86.00	12	13	3	8	8.94	1014	0.7398
12	59.06	52.94	86.06	9	10	3	7	6.08	1114	0.7376
13	62.00	54.14	88.95	10	10	4	9	6.00	1114	0.7267
14	62.18	56.36	92.06	10	8	4	9	5.36	1114	0.7502
P. muricata										
1B	58.25	49.01	78.86	10	9	3	8	5.19	1113	0.6763
2	58.70	49.32	79.16	9	8	4	8	5.62	1113	0.6740
3	59.21	53.15	80.36	10	9	3	8	5.21	1113	0.6762
4V	56.78	49.70	79.90	11	11	3	8	6.48	1114	0.7089
5	69.38	56.69	96.49	13	10	4	10	7.05	1213	0.7058
6	71.09	55.25	93.89	8	7	6	9	4.99	1213	0.6568
7	56.51	48.08	80.09	13	14	3	9	8.17	1114	0.7184
8	67.28	57.00	93.40	10	9	5	9	5.92	1113	0.6993
9	55.04	51.80	81.62	9	10	3	7	4.99	1014	0.7469
10	55.85	47.57	78.77	10	10	3	7	5.48	1114	0.7137
11	69.98	66.40	99.80	12	9	5	10	6.58	1014	0.7147
12	63.94	57.60	95.10	10	9	5	9	5.70	1114	0.7552
13	68.60	59.60	96.00	13	10	4	10	7.65	1113	0.7045
14	65.87	61.40	94.10	10	8	5	9	5.62	1014	0.7171
15	73.19	62.60	101.50	12	10	6	11	7.30	1113	0.6977
19	63.68	59.88	88.64	13	12	4	10	7.96	1013	0.6960
20	66.64	59.60	101.75	13	11	4	10	7.24	1115	0.7825
P. murrayana										
2	61.32	50.39	82.29	11	8	3	8	4.80	1213	0.6693
5	54.59	45.05	78.50	11	8	3	8	4.36	1214	0.7368
6B	50.45	39.45	68.63	9	10	2	6	4.99	1213	0.6872
7	60.81	45.76	79.88	14	12	3	9	7.18	1313	0.6534
8N	49.64	41.78	60.64	8	11	2	6	5.61	1114	0.7236
9	59.93	52.10	79.52	12	13	3	9	7.91	1113	0.6573
10	58.70	48.83	82.23	9	9	4	8	5.51	1214	0.7098
11	55.28	46.40	73.10	10	10	3	7	5.86	1113	0.6546
12	62.02	51.34	84.68	9	7	5	9	4.54	1213	0.6855
15	56.60	45.50	80.83	7	8	4	7	4.50	1214	0.7284
16	60.50	48.26	80.30	8	9	4	8	5.27	1213	0.6975
17	59.90	48.98	83.90	6	7	4	7	3.97	1214	0.7118
18	53.72	41.84	75.65	10	11	2	7	5.94	1214	0.7247
19	55.28	43.52	76.34	8	9	3	7	5.16	1213	0.7022
20	54.86	43.22	69.98	9	10	3	7	5.78	1212	0.6208

APPENDIX TABLE 1—*Continued*

Species and sample number	X_1 (in μ)	X_2 (in μ)	X_3 (in μ)	Class number	C (in percent)	Range Min. (in code)	Range Max. (in code)	SD (in μ)	Ratio index	Cos A
21	63.02	53.42	89.24	12	10	3	9	6.54	1114	0.7180
22	54.62	48.92	79.64	7	7	3	6	4.16	1114	0.7382
23	60.74	45.98	80.18	8	9	4	8	5.29	1313	0.6583
27A	55.28	38.06	70.70	11	9	3	8	5.03	1412	0.6301
27B	58.40	40.52	73.82	13	9	2	8	5.08	1412	0.6153
28	60.80	46.40	81.82	9	8	4	8	5.14	1313	0.6784
30	57.68	46.76	80.84	10	10	3	7	5.63	1214	0.7136
31	63.82	52.70	86.90	7	7	5	8	4.29	1213	0.6830
34	52.76	47.84	76.82	8	9	3	6	4.53	1014	0.7355
35	54.98	49.46	79.46	7	8	4	7	4.14	1114	0.7299
36	66.44	53.05	83.06	12	10	4	10	6.63	1212	0.6013
37	54.14	45.44	73.64	11	10	3	8	7.37	1113	0.6808
38T	58.70	46.94	79.67	7	7	4	7	4.14	1213	0.6825
P. ponderosa										
1	62.27	56.97	84.83	10	9	4	9	6.14	1013	0.6791
2	53.36	51.80	75.99	10	11	3	7	5.93	1014	0.7127
3	67.64	61.49	97.99	11	10	5	10	6.48	1114	0.7409
4	63.14	55.52	91.43	13	10	3	9	6.42	1114	0.7343
6	66.20	56.12	91.61	13	11	4	10	7.03	1113	0.6963
7	68.24	61.49	94.22	16	13	3	11	9.22	1113	0.6905
9	59.04	50.55	81.15	13	12	3	9	7.07	1113	0.6892
12N	68.45	55.85	93.26	12	10	4	10	7.09	1114	0.7374
12	72.08	60.47	101.80	14	8	5	10	5.51	1114	0.7374
15	61.52	55.88	91.79	11	8	3	8	5.04	1114	0.7568
15T	64.04	57.86	93.56	7	7	5	8	4.63	1114	0.7389
16	64.43	56.69	93.56	13	7	5	9	5.00	1114	0.7368
17	66.62	63.08	92.72	10	8	5	9	5.71	1013	0.6957
18	65.78	63.50	96.22	14	11	4	11	7.27	1014	0.7335
19	61.82	54.68	85.25	11	10	4	9	6.18	1113	0.6903
21	66.38	60.75	93.62	8	7	5	9	5.20	1014	0.7076
22	61.82	58.10	86.84	12	11	4	9	6.77	1014	0.7032
23F	62.09	54.98	86.93	11	9	3	8	5.60	1114	0.7035
24	61.82	52.16	90.10	10	8	4	8	5.00	1114	0.7465
24T	60.50	50.30	87.80	9	9	4	8	5.48	1214	0.7448
25	68.00	61.04	97.91	11	10	5	10	7.02	1114	0.7269
26	66.80	59.06	98.36	10	7	5	9	4.84	1114	0.7489
27	71.48	59.48	95.35	10	10	5	10	7.49	1213	0.6630
31	60.56	55.34	90.08	8	8	4	7	4.66	1014	0.7532
32	52.70	51.00	76.38	11	14	2	7	7.45	1014	0.7262
32N	58.25	53.96	82.94	14	11	3	9	6.34	1014	0.7148
33	64.46	56.60	93.02	11	10	4	9	6.69	1114	0.7314
34	55.34	49.04	79.70	9	10	3	7	5.46	1114	0.7285
34N	58.73	53.90	84.08	13	11	3	9	6.89	1014	0.7199
35	65.00	61.76	102.80	11	11	4	9	6.99	1015	0.7999

APPENDIX TABLE 1—*Continued*

Species and sample number	X_1 (in μ)	X_2 (in μ)	X_3 (in μ)	Class number	C (in per-cent)	Range Min. (in code)	Range Max. (in code)	SD (in μ)	Ratio index	Cos A
36	56.42	54.74	84.26	10	10	3	7	5.49	1014	0.7494
37	65.12	55.40	92.42	9	8	4	8	5.42	1115	0.7197
39	53.75	52.64	84.08	8	6	4	7	3.53	1015	0.7853
40	63.65	57.02	96.92	10	8	4	9	5.18	1115	0.7775
40N	62.96	57.32	99.68	9	5	4	8	3.81	1015	0.8101
41	66.86	58.60	100.82	10	9	5	9	5.69	1115	0.7725
P. ponderosa var. *scopulorum*										
2	58.59	60.69	81.22	8	8	4	8	4.45	1213	0.6945
3	63.44	52.70	87.44	10	10	4	9	6.33	1113	0.6942
4	63.08	56.72	90.98	11	12	4	9	6.44	1114	0.7282
11*	48.05	43.96	60.98	8	10	2	6	4.78	1014	0.7550
P. ponderosa var. *washoensis*										
28	65.39	55.76	91.50	13	10	4	10	6.33	1113	0.7062
29	64.96	58.60	90.80	13	11	4	10	7.41	1113	0.7009
29B	65.48	53.72	92.90	13	9	3	9	5.79	1214	0.7242
30	62.36	55.25	88.10	10	8	4	8	5.28	1114	0.7114
P. radiata										
1	65.63	53.54	85.43	16	16	3	10	10.25	1213	0.6403
2	58.10	51.41	89.12	11	10	3	8	6.68	1115	0.7868
3B	60.17	58.82	94.10	10	10	4	9	6.12	1015	0.7854
5	67.52	58.70	94.28	13	10	3	9	6.46	1113	0.7025
6	64.58	55.10	90.04	12	10	3	9	6.78	1113	0.7027
7	60.20	54.08	88.16	11	10	3	8	6.10	1114	0.7417
P. remorata										
3	58.01	51.05	76.61	8	8	4	7	4.86	1113	0.6533
4	63.80	55.40	88.76	8	6	5	9	3.88	1113	0.6993
5	54.32	53.30	83.72	9	7	3	7	3.93	1015	0.7731
6	63.08	54.20	86.60	8	9	5	8	5.61	1113	0.6880
P. sabiniana										
20	68.33	55.94	94.94	16	13	3	11	8.72	1213	0.7036
3	63.50	52.64	94.37	10	10	4	9	6.29	1214	0.7694
3T	63.64	51.80	93.98	9	9	5	9	5.91	1214	0.7668
4	77.87	63.68	105.50	14	8	5	12	6.40	1213	0.6789
5*	47.78	40.40	66.38	10	10	1	5	4.93	1113	0.7307
6	72.98	60.38	100.34	11	10	5	10	7.02	1213	0.6922
7	65.42	60.72	95.36	11	12	4	9	8.00	1014	0.7340
8	61.16	47.30	85.28	9	9	4	8	5.56	1213	0.7151
9	64.16	60.44	101.06	12	11	4	10	7.20	1015	0.7981

APPENDIX TABLE 1—*Continued*

Species and sample number	X_1 (in μ)	X_2 (in μ)	X_3 (in μ)	Class number	C (in percent)	Range Min. (in code)	Range Max. (in code)	SD (in μ)	Ratio index	Cos A
10	66.56	56.84	92.54	10	9	5	9	6.01	1113	0.7000
11	66.85	58.70	97.73	9	8	5	9	5.64	1114	0.7433
14	63.65	58.28	97.94	9	8	5	9	5.17	1015	0.7841
15	70.52	62.78	93.56	13	11	4	10	7.80	1113	0.6573
16	67.18	58.54	93.32	14	11	5	10	7.63	1113	0.6976
17	78.53	66.14	109.34	14	11	5	11	3.28	1113	0.7027
18	71.45	64.00	104.20	13	11	4	10	7.77	1114	0.7384
P. torreyana										
2	57.57	49.54	78.18	11	10	3	8	5.84	1113	0.6779
4	62.96	58.34	89.30	9	7	4	8	4.39	1014	0.7116
5	66.65	59.72	89.24	10	9	5	9	6.07	1113	0.6650
6T	65.09	58.10	88.58	11	8	4	9	5.65	1113	0.6786

APPENDIX II

FREQUENCY SPECTRA OF BREADTH OF CORPUS FROM POLLEN GRAINS OF SAMPLES OF *Pinus* SPECIES

The species and sample numbers listed include all those whose measurements of corpus of pollen grains I took. The list also includes a few samples from other sources, samples measured from old slides deposited in other institutions, and some incomplete samples lacking measurements of the breadth of saccus and/or of overall size. These samples, not used in computational programing, are marked by an asterisk. Samples with data that are in fact complete but, for various reasons, have not been used in computational programs, are marked by two asterisks.

Appendix table 2 is arranged alphabetically by species. Sample size is 100 unless otherwise indicated. The abbreviations for taxa are as follows:

A × R	*P. attenuradiata,* a hybrid
J × C	Hybrid of *P. jeffreyi* and *P. ponderosa*
J × C	Hybrid of *P. jeffreyi* and *P. coulteri*
L	Herbarium specimens labeled as *P. latifolia* or as a variety of either *P. contorta* or *P. murrayana* (all listed under *P. contorta* var. *latifolia* in appendix table 1)
PS	*Pinus ponderosa* var. *scopulorum*
PW	*Pinus ponderosa* var. *washoensis*
Z	*Pinus arizonica*

The table gives spectra for measurements only of the breadth of corpus. The class frequencies are entered under each class midpoint, which are shown in code in the upper row and in μ in the lower row. I used the coded midpoints in spectrum-fitting by coarse grouping as well as in computational programing (see Appendix III).

APPENDIX TABLE 2

FREQUENCY SPECTRA OF BREADTH OF CORPUS FROM POLLEN GRAINS OF SAMPLES OF *Pinus* SPECIES

(N = 100, except as noted)

Species and sample number	N	X_1 (in μ)	Pollen frequency by class midpoint (Coarse grouping in upper row in code, fine grouping in lower row in μ): 1		2		3		4		5		6		7		8		9		10		11		12		13		14		15	
			29	32	35	38	41	44	47	50	53	56	59	62	65	68	71	74	71	80	83	86	89	92	95	98	101	104	107	110	113	116
P. albicaulis																																
1	..	55.58	..	..	..	4	6	6	6	11	13	4	14	21	4	6	5	..	..	..	..	..	..	..	..	..	..	..	..	..	..	..
2	..	55.70	..	..	..	..	..	..	5	15	25	22	12	15	3	3	..	..	..	..	..	..	..	..	..	..	..	..	..	..	..	..
7	..	58.16	..	..	..	2	2	8	6	8	16	6	6	6	22	8	6	0	2	2	..	..	..	..	..	..	..	..	..	..	..	..
9	..	60.68	..	..	..	..	..	..	..	4	20	14	8	14	24	6	8	2	..	..	..	..	..	..	..	..	..	..	..	..	..	..
10	..	55.64	..	..	..	..	..	..	2	14	36	14	16	12	6	..	..	..	..	..	..	..	..	..	..	..	..	..	..	..	..	..
11	..	60.11	..	..	..	..	1	0	2	6	18	7	7	28	18	7	3	3	..	..	..	..	..	..	..	..	..	..	..	..	..	..
13	..	64.31	..	..	..	..	..	..	..	1	4	5	9	25	27	13	6	9	1	..	..	..	..	..	..	..	..	..	..	..	..	..
14b	..	60.92	..	..	..	..	..	..	..	8	12	6	24	12	20	10	4	4	..	..	..	..	..	..	..	..	..	..	..	..	..	..
B7	..	53.60	..	..	..	2	2	4	10	18	22	18	8	10	6	..	..	..	..	..	..	..	..	..	..	..	..	..	..	..	..	..
P. aristota																																
1	..	45.05	..	2	6	9	15	19	18	18	12	1	..	..	..	..	..	..	..	..	..	..	..	..	..	..	..	..	..	..	..	..
2	..	48.11	..	..	1	3	11	20	12	21	27	3	2	..	..	..	..	..	..	..	..	..	..	..	..	..	..	..	..	..	..	..
3	..	52.91	..	..	1	1	4	3	6	17	35	16	8	6	3	..	..	..	..	..	..	..	..	..	..	..	..	..	..	..	..	..
3N	..	49.77	..	..	..	3	12	6	16	22	25	14	2	..	..	..	..	..	..	..	..	..	..	..	..	..	..	..	..	..	..	..
6A	..	54.38	..	1	0	2	3	4	3	13	25	19	13	11	3	2	1	..	..	..	..	..	..	..	..	..	..	..	..	..	..	..
6B	..	53.69	..	..	..	2	5	8	2	10	32	8	15	12	5	1	..	..	..	..	..	..	..	..	..	..	..	..	..	..	..	..
7	..	53.00	..	..	..	..	2	2	14	26	22	14	8	10	2	..	..	..	..	..	..	..	..	..	..	..	..	..	..	..	..	..
8	..	50.24	..	..	2	0	8	12	10	22	28	10	6	0	2	..	..	..	..	..	..	..	..	..	..	..	..	..	..	..	..	..
9	..	48.20	..	..	..	6	16	12	4	34	20	6	0	2	..	..	..	..	..	..	..	..	..	..	..	..	..	..	..	..	..	..
10	..	47.06	..	..	1	6	13	14	28	21	8	9	..	..	..	..	..	..	..	..	..	..	..	..	..	..	..	..	..	..	..	..
14	..	54.08	..	..	..	..	..	4	2	22	24	32	9	4	3	..	..	..	..	..	..	..	..	..	..	..	..	..	..	..	..	..
15	..	50.18	..	..	..	5	2	7	19	30	21	8	4	4	..	..	..	..	..	..	..	..	..	..	..	..	..	..	..	..	..	..
16	..	54.52	..	..	..	..	1	0	6	18	27	29	10	5	4	1	..	..	..	..	..	..	..	..	..	..	..	..	..	..	..	..
17	..	49.58	..	..	..	3	3	6	19	41	20	6	2	..	..	..	..	..	..	..	..	..	..	..	..	..	..	..	..	..	..	..
P. attenuata																																
1	..	59.24	..	..	..	..	..	..	..	1	6	19	13	19	22	8	8	2	1	1	..	..	..	..	..	..	..	..	..	..	..	..
2	..	58.34	..	..	..	..	..	2	2	6	17	16	15	21	12	4	5	1	..	..	..	..	..	..	..	..	..	..	..	..	..	..
3	..	57.41	..	..	..	..	..	1	2	4	32	15	13	19	11	2	1	..	..	..	..	..	..	..	..	..	..	..	..	..	..	..
4A	..	66.17	..	..	..	..	..	..	..	2	2	2	6	10	39	17	10	8	6	..	..	..	..	..	..	..	..	..	..	..	..	..
4D	**400**	**67.85**	..	..	..	..	..	..	..	2	2	7	14	49	114	62	65	42	28	10	2	0	2	1	..	..	..	..	..	..	..	..

4E	..	68.72	..	..	..	..	..	..	..	..	..	1	2	9	28	21	13	14	9	3	..	..	..	..	..	..	..	..	..	..	..	..
7	..	64.07	..	..	..	..	..	..	1	2	13	10	8	15	5	8	14	5	5	1	1	1	0	1	..	..	..	..	..	..	..	..
9 (A × R)	..	64.94	..	..	..	..	..	..	..	..	3	3	5	27	27	19	11	5	..	..	..	..	..	..	..	..	..	..	..	..	..	..
10	..	59.93	..	..	..	..	2	0	0	1	15	15	20	17	22	5	0	3	..	..	..	..	..	..	..	..	..	..	..	..	..	..
11	..	61.91	..	..	..	..	..	..	..	3	6	7	10	32	31	7	2	..	..	..	..	..	..	..	..	..	..	..	..	..	..	..
12	..	64.46	..	..	..	..	..	..	..	1	1	4	10	25	32	12	10	4	1	..	..	..	..	..	..	..	..	..	..	..	..	..
13	..	64.19	..	..	..	..	..	1	1	3	7	3	6	22	18	20	10	6	2	0	1	..	..	..	..	..	..	..	..	..	..	..
14	..	59.48	..	..	..	..	1	0	1	4	13	20	10	28	14	8	1	..	..	..	..	..	..	..	..	..	..	..	..	..	..	..
16	..	59.57	..	..	..	..	1	1	5	6	14	11	8	25	18	3	5	1	2	..	..	..	..	..	..	..	..	..	..	..	..	..
17	..	62.42	..	..	..	..	..	..	..	2	6	14	9	34	15	8	5	5	1	1	..	..	..	..	..	..	..	..	..	..	..	..
18	..	63.74	..	..	..	..	..	..	..	..	5	4	8	29	32	14	6	0	1	1	..	..	..	..	..	..	..	..	..	..	..	..
19	..	63.89	..	..	..	..	..	..	..	..	6	9	6	31	17	14	9	7	1	..	..	..	..	..	..	..	..	..	..	..	..	..
20	..	58.52	..	..	..	..	..	1	2	4	18	17	19	23	12	1	3	..	..	..	..	..	..	..	..	..	..	..	..	..	..	..
21	..	57.29	..	..	..	..	1	2	3	7	23	22	13	11	14	3	2	..	..	..	..	..	..	..	..	..	..	..	..	..	..	..
22	..	63.23	..	..	..	..	..	..	1	3	8	8	5	19	32	11	7	3	2	1	..	..	..	..	..	..	..	..	..	..	..	..
23	..	60.68	..	..	..	..	2	1	0	1	12	15	12	21	21	8	7	..	..	..	..	..	..	..	..	..	..	..	..	..	..	..
24	..	56.96	..	..	..	1	3	0	2	4	27	14	18	21	7	2	1	..	..	..	..	..	..	..	..	..	..	..	..	..	..	..
25	50	69.14	..	..	..	..	..	..	..	..	..	2	1	7	8	4	11	8	7	1	..	..	..	..	..	..	..	..	..	..	..	..
26	51	69.76	..	..	..	..	..	..	..	..	..	2	2	2	10	8	9	6	11	1	..	..	..	..	..	..	..	..	..	..	..	..
27	50	67.70	..	..	..	..	..	..	..	..	..	3	1	5	12	12	8	5	3	0	1	..	..	..	..	..	..	..	..	..	..	..
28	..	62.78	..	..	..	..	..	..	..	4	6	6	14	14	40	10	0	4	2	..	..	..	..	..	..	..	..	..	..	..	..	..
29	..	64.34	..	..	..	4	0	0	4	4	8	4	0	12	16	18	14	6	4	4	0	2	..	..	..	..	..	..	..	..	..	..
29A	..	66.05	..	..	..	1	0	0	1	0	2	6	5	9	29	18	17	5	7	..	..	..	..	..	..	..	..	..	..	..	..	..
29N	..	69.00	..	..	..	..	..	..	..	2	2	4	2	8	30	10	12	10	10	4	4	2	2	..	..	..	..	..	..	..	..	..
30	..	61.64	..	..	..	..	..	..	2	4	10	8	10	30	18	10	4	2	0	2	..	..	..	..	..	..	..	..	..	..	..	..
30T	60	59.45	..	..	..	..	..	..	..	1	10	10	13	14	10	0	2	..	..	..	..	..	..	..	..	..	..	..	..	..	..	..
31	..	69.47	..	..	..	..	..	..	1	0	0	0	3	10	17	19	18	18	11	1	1	1	..	..	..	..	..	..	..	..	..	..
32	51	58.64	..	..	..	..	..	1	1	3	4	12	11	8	9	2	..	..	..	..	..	..	..	..	..	..	..	..	..	..	..	..
33	50	64.10	..	..	..	..	..	..	..	..	6	2	2	6	16	12	5	1	..	..	..	..	..	..	..	..	..	..	..	..	..	..
34	51	70.59	..	..	..	..	..	..	..	..	..	1	2	2	11	6	8	7	11	2	0	1	..	..	..	..	..	..	..	..	..	..
35	50	68.48	..	..	..	..	..	..	..	..	2	3	0	2	13	9	7	7	4	2	0	1	..	..	..	..	..	..	..	..	..	..
36	50	66.08	..	..	..	..	..	..	..	..	2	1	2	9	17	6	5	7	0	1	..	..	..	..	..	..	..	..	..	..	..	..
P. balfouriana																																
1	..	56.72	..	..	..	..	1	3	2	10	24	12	21	18	2	3	2	1	..	..	..	..	..	..	..	..	..	..	..	..	..	..
2	..	50.28	..	..	1	0	6	11	18	17	31	10	3	2	1	..	..	..	..	..	..	..	..	..	..	..	..	..	..	..	..	..
4A	..	50.96	..	..	..	2	7	14	7	21	27	8	4	9	1	..	..	..	..	..	..	..	..	..	..	..	..	..	..	..	..	..
4H	..	51.32	..	..	..	..	7	7	12	25	30	6	5	7	0	1	..	..	..	..	..	..	..	..	..	..	..	..	..	..	..	..
5	..	58.70	..	..	..	..	2	2	4	4	20	12	12	16	16	10	0	0	0	2	..	..	..	..	..	..	..	..	..	..	..	..
6	50	51.44	..	..	..	1	3	2	3	10	21	8	1	1	..	..	..	..	..	..	..	..	..	..	..	..	..	..	..	..	..	..
8	..	55.52	..	..	..	..	2	0	5	10	33	15	13	17	5	..	..	..	..	..	..	..	..	..	..	..	..	..	..	..	..	..
9	50	53.72	..	..	..	1	1	1	2	13	12	7	7	5	0	1	..	..	..	..	..	..	..	..	..	..	..	..	..	..	..	..
11	34	50.26	..	1	2	0	0	2	0	9	14	6	..	..	..	..	..	..	..	..	..	..	..	..	..	..	..	..	..	..	..	..
18	..	54.23	..	..	..	1	2	1	13	17	14	24	16	4	8	..	..	..	..	..	..	..	..	..	..	..	..	..	..	..	..	..

APPENDIX TABLE 2—*Continued*

Species and sample number	N	X_1 (in μ)	Pollen frequency by class midpoint (Coarse grouping in upper row in code, fine grouping in lower row in μ) 1		2		3		4		5		6		7		8		9		10		11		12		13		14		15	
			29	32	35	38	41	44	47	50	53	56	59	62	65	68	71	74	71	80	83	86	89	92	95	98	101	104	107	110	113	116
P. contorta																																
1	..	52.43	..	..	..	..	..	1	9	28	39	16	7	..	..	..	..	..	..	..	..	..	..	..	..	..	..	..	..	..	..	..
2	..	58.52	..	..	..	..	..	..	1	2	17	12	43	19	5	1	..	..	..	..	..	..	..	..	..	..	..	..	..	..	..	..
5	..	61.91	..	..	..	..	..	..	1	1	10	12	11	23	28	5	5	2	2	..	..	..	..	..	..	..	..	..	..	..	..	..
6	50	52.36	..	..	..	..	1	2	4	12	23	10	4	..	..	..	..	..	..	..	..	..	..	..	..	..	..	..	..	..	..	..
7A	300	58.71	..	..	..	..	1	3	0	9	74	43	49	69	28	15	7	0	2	..	..	..	..	..	..	..	..	..	..	..	..	..
11	50	52.82	..	..	..	..	1	0	2	10	26	9	1	0	1	..	..	..	..	..	..	..	..	..	..	..	..	..	..	..	..	..
12	50	51.14	..	..	..	3	0	1	11	13	4	16	1	0	1	..	..	..	..	..	..	..	..	..	..	..	..	..	..	..	..	..
14	51	56.42	..	..	..	..	..	1	2	1	13	17	9	4	3	0	1	..	..	..	..	..	..	..	..	..	..	..	..	..	..	..
15	50	48.44	..	..	..	..	5	6	19	6	10	3	0	1	..	..	..	..	..	..	..	..	..	..	..	..	..	..	..	..	..	..
16	50	49.04	..	..	1	2	7	2	5	14	13	4	2	..	..	..	..	..	..	..	..	..	..	..	..	..	..	..	..	..	..	..
17	50	56.24	..	..	..	..	..	..	3	3	14	12	6	10	1	1	..	..	..	..	..	..	..	..	..	..	..	..	..	..	..	..
18	..	56.72	..	..	..	..	..	..	..	4	30	32	12	16	6	..	..	..	..	..	..	..	..	..	..	..	..	..	..	..	..	..
19	55	51.91	..	..	..	..	1	4	12	11	10	12	0	2	2	1	..	..	..	..	..	..	..	..	..	..	..	..	..	..	..	..
20	55	51.20	..	..	..	..	1	1	8	18	15	12	..	..	..	..	..	..	..	..	..	..	..	..	..	..	..	..	..	..	..	..
22	105	57.11	..	..	..	..	..	1	1	7	37	13	14	17	11	4	..	..	..	..	..	..	..	..	..	..	..	..	..	..	..	..
22A	50	56.42	..	..	..	..	..	1	1	2	21	7	6	6	3	3	..	..	..	..	..	..	..	..	..	..	..	..	..	..	..	..
22B	55	57.74	..	..	..	..	..	..	..	5	16	6	8	11	8	1	..	..	..	..	..	..	..	..	..	..	..	..	..	..	..	..
24	50	51.74	..	..	..	1	1	3	4	9	24	4	2	2	..	..	..	..	..	..	..	..	..	..	..	..	..	..	..	..	..	..
25	50	54.68	..	..	..	..	1	1	1	5	26	4	2	6	4	..	..	..	..	..	..	..	..	..	..	..	..	..	..	..	..	..
26	50	44.96	..	..	..	4	13	11	10	9	3	..	..	..	..	..	..	..	..	..	..	..	..	..	..	..	..	..	..	..	..	..
27	50	51.80	..	..	..	..	3	5	4	5	15	16	2	..	..	..	..	..	..	..	..	..	..	..	..	..	..	..	..	..	..	..
8* (L)	..	53.69	..	..	..	..	..	4	4	8	53	11	10	10	..	..	..	..	..	..	..	..	..	..	..	..	..	..	..	..	..	..
9* (L)	..	57.80	..	..	..	..	..	..	4	7	13	23	18	26	7	2	..	..	..	..	..	..	..	..	..	..	..	..	..	..	..	..
23* (L)	..	50.18	..	..	..	..	2	6	8	16	24	32	6	4	2	..	..	..	..	..	..	..	..	..	..	..	..	..	..	..	..	..
P. contorta var. *bolanderi*																																
1	..	49.67	..	..	1	2	4	21	14	25	18	4	3	5	3	..	..	..	..	..	..	..	..	..	..	..	..	..	..	..	..	..
2B	..	46.81	..	..	1	4	32	21	17	12	10	0	1	2	..	..	..	..	..	..	..	..	..	..	..	..	..	..	..	..	..	..
3	..	51.89	..	..	..	1	2	8	7	23	37	10	4	5	2	..	..	..	..	..	..	..	..	..	..	..	..	..	..	..	..	..
4	..	48.26	..	1	0	4	17	10	13	26	17	8	2	1	1	..	..	..	..	..	..	..	..	..	..	..	..	..	..	..	..	..
5	50	52.28	..	..	..	..	..	2	6	9	23	6	3	1	..	..	..	..	..	..	..	..	..	..	..	..	..	..	..	..	..	..
6	50	52.46	..	..	..	..	3	0	1	13	24	3	3	3	..	..	..	..	..	..	..	..	..	..	..	..	..	..	..	..	..	..
7	50	49.88	..	..	..	..	4	3	8	17	13	4	1	..	..	..	..	..	..	..	..	..	..	..	..	..	..	..	..	..	..	..

APPENDIX TABLE 5

SUCCESSIVE CLASSIFICATIONS OF *Pinus* BY ENTERING VARIABLES IN DIFFERENT SEQUENCES AND NUMBER

Species	Number of samples	Successive classifications: Program III, Trial 1	2	3	Program IV, Trial 4	5	6	Optimum result	Percent classified
Group A									
P. albicaulis	8	5	6	6	6	5	7	7	87.5
P. flexilis	6	4	3	2	2	4	2	4	66.7
P. monticola	11	7	6	7	7	7	7	7	63.6
P. monophylla	7	7	7	7	7	7	7	7	100.0
P. lambertiana	9	8	8	8	8	8	8	8	89.0
P. aristata	13	9	8	8	9	9	8	9	70.0
P. balfouriana	8	6	7	7	7	6	7	7	87.5
P. torreyana	3	3	3	3	3	3	3	3	100.0
Total	65	49	48	48	49	49	49	52	80.0

Species	Number of samples	Successive classifications: Program III, Trial 7	8	Program IV, Trial 9	10	11	12	Optimum result	Percent classified
Group B									
P. contorta	23	11	11	9	10	9	8	11	47.8
P. contorta var. *bolanderi*	5	5	5	5	5	4	5	5	100.0
P. murrayana	29	15	14	13	13	14	14	15	51.7
P. coulteri	10	8	8	8	8	4	8	8	80.0
P. jeffreyi	16	13	14	14	13	8	13	14	87.5
P. ponderosa	33	7	5	7	6	10	9	10	30.3
P. attenuata	28	8	10	12	11	10	11	12	42.8
P. muricata	13	4	3	3	2	1	2	4	30.7
P. radiata	6	4	3	4	0	2	1	4	66.7
P. remorata	4	3	4	4	3	3	4	4	100.0
P. sabiniana	14	7	4	5	7	10	8	10	70.1
Total	181	85	81	84	78	75	83	97	53.6

the pattern of the cappa, as *P. ponderosa* has an unmistakable Y pattern. But the mutual misclassification among *P. attenuata, P. muricata,* and *P. sabiniana* is difficult to overcome, and would probably remain difficult under the same type of statistical analysis.

For Group *B* (appendix table 4), the optimum classification gives 97 out of 181, but with individual trials, the best one (trial 7) gives only 85 out of 181, approximately 47 per cent. Considering *P. contorta, P. contorta* var. *bolanderi,* and *P. murrayana* as a complex, the percentage would then be increased to 54 for individual trial 7, and to 61 for maximum results.

The best trial, program III, trial 7 (appendix table 4), is found by entering only six variables in the order shown in appendix table 5. It is found that variables *C* and maximum class midpoint are not useful, whereas variable *SD* is very useful.

CONCLUSION

The discriminant analysis has so far sustained the belief that pollen statistics can be useful in specific determination among species of *Pinus*. The varying degree of correct classifications, based on a different number of and different kinds of variables, makes it necessary to select suitable statistics and number of variables in order to obtain the best results for any individual taxon. The specific identification done in the manner described in Part III gives results superior to this type of analysis. The former method considers every individual example as an independent universe, whereas discriminant analysis, as it is made, groups the possible "races" within a species into one universe, thereby affecting the results.

APPENDIX IV

The Degree of Resemblance Measured by the Gamma Index

When a number of samples of one kind (taxon, species, group, and so on), *i,* is compared with a number of samples of another kind, *j,* the degree of resemblance of one to the other is expressed by an index, which is here named the gamma index (Γ). The index describes the frequency of connections between samples of the same taxon (intrataxon), and those between samples of different taxa (intertaxa). For intrataxon connections the following equation is used:

$$\Gamma_{ii} = \frac{\sum f_{ii}}{N_i(N_i - 1)/2} = I_1 , \quad (1)$$

where N_i is the total number of samples of taxon *i,* and f_{ii} is the sum of connections between these samples.

For intertaxa comparisons, when each sample of taxon *i* is compared with each sample of taxon *j,* the following equation is used:

$$\Gamma_{ij} = \frac{\sum f_{ij}}{N_i N_j} = I_2 . \quad (2)$$

In both equations the denominator is the maximum possible connection between a given N_i and N_j. The numerator is the sum of the actual connections found. Hence the index is in fact a ratio of actual occurrences against the maximum possibility.

The ratio $\Gamma_{ij} / \Gamma_{ii}$ $(=I_2 / I_{1_i})$ expresses the probability of the samples of *i* being classified as *j,* and the ratio $\Gamma_{ij} / \Gamma_{jj}$ expresses the probability of samples of *j* being classified as *i.* The composite ratio of $2\Gamma_{ij} / \Gamma_{ii}+\Gamma_{jj}$ expresses the average probability of one being mistaken for another.

To find the connections between samples, compare their statistics. These statistics, or variables, are those used in the discriminant analysis (Appendix III) with two changes in the procedure: (1) the standard deviation (*Sd*) is not employed; (2) the ratio index is used instead.

The variables are weighted according to a scheme outlined in appendix table 6. The scheme is arrived at by testing paired lots drawn from the same universe. Suitable weighting should show that the paired lots are the same, or resemble each other to a very high degree. An identity will have a score of 16 points, and any score equal to or above 14 is considered a significant indication of resemblance between two samples.

If every one of the samples of taxon *i* scores 14 to 16 points when it is compared with another sample of taxon *i* or of taxon *j,* the gamma indexes will both be approaching 1.00. This means that all samples of taxon *i* are correctly classified as taxon *i* and that all samples known as taxon *j* are identical with those of taxon *i.* Assuming samples *i* and *j* are indeed from two distinct taxa, then the complete resemblances between them indicate that statistics have failed to distinguish these two taxa. If the intertaxa gamma index is low while the intrataxon index is high, then the statistics used are capable of discriminating the two taxa.

Theoretically, taxonomic judgment needs the support of many independent variables, and pollen is only one of them. The different variables used here are all from pollen grains, and only a few of them are independent. Hence, one normally would not expect a taxonomic judgment based only on these variables. Experience, however, shows that the gamma index does in many cases serve to distinguish species from species, though the distinction is not 100 per cent. The degree of

APPENDIX TABLE 6

Scheme of Weighting the Variables Used in Computing Scores of Resemblances between Samples

Variables	Weighting scheme		Point value	Maximum possible points
X_1, X_2, X_3	If difference between means of the same kind of measurement from two samples is	less than or equal to 2.99μ	2	6
		from 3.0μ to 3.7μ	1	
		more than 3.70μ	0	
Class number, coefficient of variance, maximum and minimum class of spectrum	If difference between values of the same variable from two samples is	less than 2	1	4
		equal to or more than 2	0	
Ratio index	If difference between ratio indexes of two samples	equals 0	2	2
		equals 1 or 100	1	
		is other than those above	0	
Cos A index	If difference between the indexes of two samples is	less than or equal to 70	4	4
		between 71 and 170	3	
		between 171 and 300	2	
		between 301 and 450	1	
		more than 450	0	
The total maximum possible points				16

Note: A total of 14 points out of 16 (= 87 percent) and above is considered significant.

resemblance between taxa is not always high, neither is that between samples of one species. A ratio R_i, R_j, and R_{ij} of 70 per cent is taken as the minimum level of resemblance (i.e., $R = 0.7$).

The data of 27 pollen samples of *P. murrayana* (*i*), 4 samples of *P. contorta* var. *latifolia* (*j*), 5 samples of *P. contorta* var. *bolanderi* (*k*), and 18 samples of *P. contorta* (*p*) are used as an example to show the interpretation of the gamma indexes, their ratio, and composite probabilities. Their intrataxon connections (*ii, jj, kk, pp*) and their intertaxa connections (*ij, ik, ip, jk, jp, kp*) are both termed frequencies (*f*). These are obtained from sample-to-sample comparisons. Pairs of samples, whether of the same species or from different taxa and having 14 points or more in common, are recorded and counted as one connection (*f*) between the two individuals.

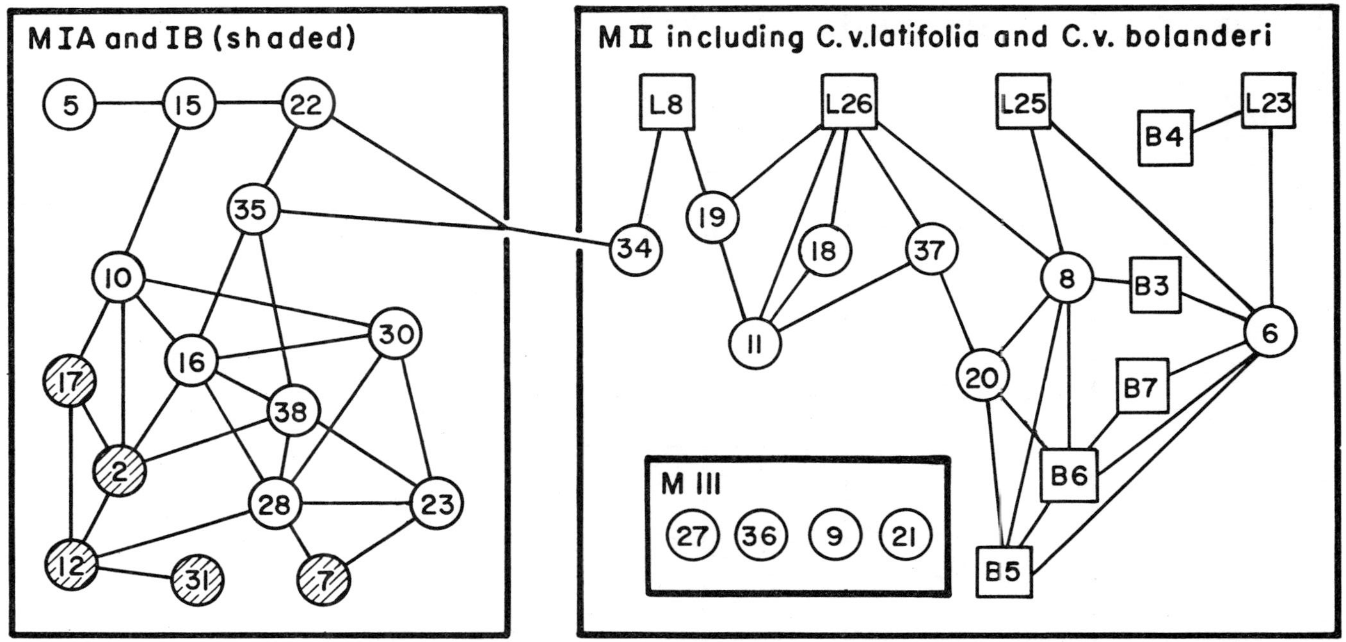

Appendix fig. 1. Diagram showing the connections between individual samples as expressed by gamma indexes. Samples (numbered) of *Pinus murrayana* (in circles). *P. contorta* var. *latifolia* (*L.* in squares), and *P. contorta* var. *bolanderi* (*B.* in squares) are connected when an index of 14 points or more has been scored. The connections with *P. contorta* are shown in appendix figure 2.

Note the subdivisions of *P. murrayana* into three groups (MIA and IB, MII, MIII). The shaded samples are the subgroup MIB. This subgroup has no connection with any of the other taxa.

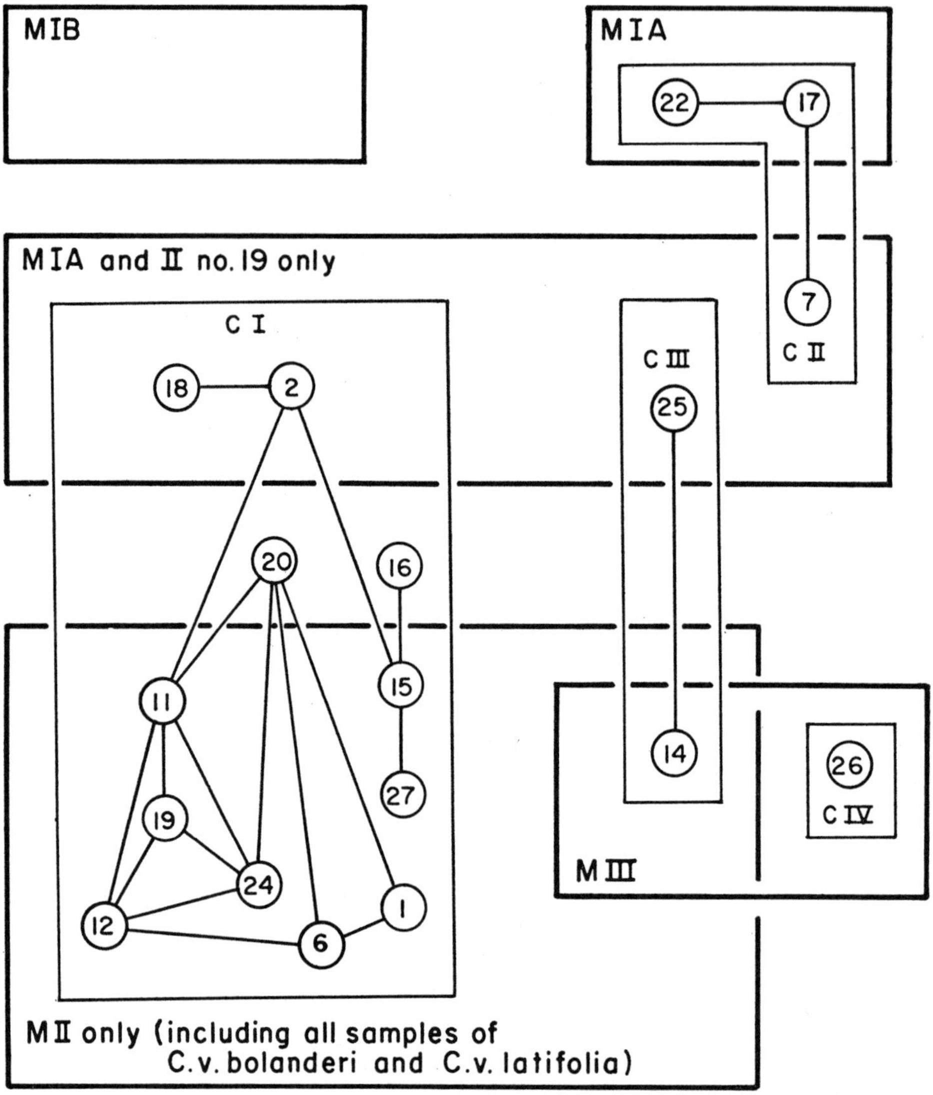

Appendix fig. 2. Diagram showing connections between pollen groups of *P. contorta* and *P. murrayana*. The groups or sets of *P. murrayana* (enclosed in heavy lines) (MIA, MIB, MII, MIII) overlap those of *P. contorta* (enclosed in thin lines) (CI, CII, CIII, CIV). The *P. contorta* group CIII is also connected with the varieties *bolanderi* and *latifolia*. The fact that *P. murrayana* group MIB remains isolated indicates the justification of considering it a separate species.

APPENDIX TABLE 7

Scores, Indexes, Ratios and Composite Probabilities of the *P. contorta* Complex

N = number of samples			27	4	5	18
Taxon			i = *P. murrayana*	j = *P. contorta* var. *latifolia*	k = *P. contorta* var. *bolanderi*	p = *P. contorta*
Frequency = f		i	35	10	9	53
(number of samples connected by a score of 14 or more)		j		0	1	11
		k			2	12
		p				20
Gamma index	$I_1 = 2f/N(N-1) \times 100$		$I_{ii} = 10.0$	$I_{jj} = 0$	$I_{kk} = 20.0$	$I_{pp} = 13.1$
	$I_2 = f/N_1N_2 \times 100$	i		$I_{ij} = 9.2$	$I_{ik} = 6.7$	$I_{ip} = 10.9$
		j			$I_{jk} = 5.0$	$I_{jp} = 15.3$
		k				$I_{kp} = 13.3$
Ratio (R)	R I_2/I_1					
	R of ii			$I_{ij}/I_{ii} = 0.92$	$I_{ik}/I_{ii} = 0.67$	$I_{ik}/I_{ii} = 1.09$
	R of ji					
	R of kk		$I_{ik}/I_{kk} = 0.33$	$I_{jk}/I_{kk} = 0.25$		
	R of pp		$I_{ik}/I_{pp} = 0.83$	$I_{jp}/I_{pp} = 1.17$	$I_{kp}/I_{pp} = 1.01$	$I_{kk}/I_{kk} = 0.67$
Composite probability	A = ii			$I_{ij}/I_a + I_b = 184.0$	$I_{ik}/I_a + I_b = 44.6$	$I_{ip}/I_a + I_b = 94.4$
	A = jj				$I_{jk}/I_a + I_b = 50.0$	$I_{jp}/I_a + I_b = 216.0$
	A = kk					$I_{kp}/I_a + I_b = 80.4$

With the values N and f, the gamma index can be obtained by substitution into appendix equations (1) and (2). The ratio (R) and the probabilities (P_c) are obtained from the indexes by setting $R = I_2/I_1$ and $P_c = 2I_2/(I_{1a} + I_{1b})$, where, if $a = ii$, $b = jj$, kk, or pp; if $a = jj$, $b = kk$ or pp; and if $a = kk$, $b = pp$. The results are shown in appendix table 7. The actual sample-to-sample connections are rendered into two diagrams (appendix figs. 1, 2).

The composite probability between taxa indicates that (1) samples of *murrayana* very much resemble those of *latifolia* (184.0 per cent) and those of *contorta* (94.4 per cent), but are not the same as those of *bolanderi* (44.6 per cent); (2) samples of *contorta* very strongly resemble those of *latifolia* (216.0 per cent) and those of *bolanderi* (80.4 per cent; and (3) samples of *latifolia* are not the same as those of *bolanderi* (50.0 per cent).

The conclusion is that to differentiate the pollen samples of *P. murrayana* from those of *P. contorta* var. *bolanderi* is possible, but to distinguish *P. contorta* var. *latifolia* from either *P. contorta* or *P murrayana* is not possible. Furthermore, the difference between pollen samples of *P. contorta* and *P. murrayana* is also very small.

From the diagrams (appendix figs. 1, 2), subdivisions of the samples into groups and subgroups can be made. Notice that while the samples of these four taxa overlap one another, a group of *P. murrayana* samples (*IB*) (appendix fig. 1) remains isolated with no intergroup connections. There is therefore some justification in considering *P. murrayana* and *P. contorta* as distinct species. The fact that all the samples of group *IB* came from localities in the central Sierra Nevada might imply some geographical differences.

The discussion here is intended only to show the usefulness of the gamma indexes, and, since, the work is still in progress, the conclusions should not be taken as final. Data on gamma indexes of all 18 species studied have been reported (Ting, 1965). Suffice it to state here that these indexes give much the same results as those obtained from the discriminant analysis discussed in Appendix III.

APPENDIX V

Comparison of Sizes of Fresh and Acetolyzed Pollen Grains of *Pinus*

Date: January to May, 1960.

Material: Three samples each of pollen grains of *Pinus jeffreyi* and *P. muricata* were used in this experiment. They are *P. jeffreyi* 10, 26, 27, and *P. muricata* 11, 12, 13. *P. jeffreyi* 26 and 27 were collected from two separate microsporangiate strobili of the same tree, 12 miles west of Frazer Park, Ventura County, California; and sample 10 was fresh pollen mixed from four trees, supplied by the Institute of Forest Genetics, Placerville. *P. muricata* 11 was from Mendocino, 12 was from Monterey, and 13 was from San Vincente, Baja California.

Treatment: Fresh desiccated and acetolyzed pollen grains were used in the test. The pollen grains were desiccated either by washing with xylene and alcohol, or by placing in an oven at a temperature of 150° or 300° F for 10 or 15 minutes.

Mounting of slides: Slides were mounted in glycerine jelly. The "dry" mounts are slides of desiccated pollen grains mounted by dropping melted glycerine jelly over the pollen grains. The "wet" mounts have a drop of water added to the desiccated pollen grains before dropping on the melted glycerine jelly. Exposure to heat sufficient only to melt the jelly is permitted. "Heated" slides were made by heating the mounted slide over an alcoholic lamp until the jelly bubbled.

Measurements: Measurements were made immediately after the slides were ready for observation. The dimensions measured are breadth of corpus ($\overline{X}_1$), breadth of saccus ($\overline{X}_2$), and the overall breadth ($\overline{X}_3$). I used $\overline{X}_1$ for comparison in most of my tests. Differences between values are usually expressed as increments in percentages of the smaller value. For easy comparison with data of other workers, I used a proportion when necessary. A proportion is equal to 100 plus the percentage. Two times the standard error is given ($2SE$), for this is approximately equal to the limits at the 5 per cent level of confidence.

Results (appendix tables 8*a*, *b*):

1. Untreated, desiccated pollen grains in dry mount, heated, induced an immediate swelling of the pollen grains. The difference in size lies between 14 per cent and 23 per cent for $\overline{X}_1$; is 70 per cent for $\overline{X}_2$; and is 39 per cent for $\overline{X}_3$. The differences are significant (appendix table 8*a*, *P. jeffreyi* 27, slide 1*a* compared with 1*b*; *P. jeffreyi* 26, slide 2*a* compared with 2*b*, 3*a* with 3*b*).

2. Size of pollen, mounted dry and heated, but not acetolyzed, is not significantly different from size when acetolyzed. The maximum difference is about 4 per cent (77.22μ and 74.30μ, respectively, in appendix table 8*a*, *P. jeffreyi* 26, slides 3*aa, b;* see also *P. jeffreyi* 27, slide 1*b* compared with 2; *P. jeffreyi* 26, slide 2*b* compared with slides 1*a*, *b*, *c*, and slide 2).

3. Mounting procedures make no difference in pollen size, if the pollen sample was acetolyzed (appendix table 8*a*, *P. jeffreyi* 27, slide 2, lots 1, 2; *P. jeffreyi* 26, slides 1*a*, *b*, *c*, 2, 3*aa; P. jeffreyi* 10, slides 2*a*, *b*, *N*).

4. Differences in size of saccus ($\overline{X}_2$) and overall size ($\overline{X}_3$) do exist in the comparisons made in (2) and (3), above. These differences are probably not caused by treatment. The mean of the corpus ($\overline{X}_1$) is therefore regarded as a better criterion for reliability in identification purposes.

5. There is no significant difference in size of $\overline{X}_1$ of pollen grains from two different strobili of the same tree when acetolyzed. But their original nontreated sizes are significantly different (dry mount, nontreated slide 1*a* of *P. jeffreyi* 27 and dry mount, nontreated slide 2*a* of *P. jeffreyi* 26 in appendix table 8a).

APPENDIX TABLE 8*a*

VARIATION IN SIZE OF *Pinus jeffreyi* POLLEN WITH DIFFERENT TREATMENT AND MOUNTING PROCEDURES

Species and sample number	Slide number	Treatment: Mounting	Treatment: Chemical	N	X_1 (in μ)	2SE (in μ)	X_2 (in μ)	2SE (in μ)	X_3 (in μ)	2SE (in μ)
P. jeffreyi 27	1*a*	Dry.........	None	40	66.72	2.65	.....	....	......	....
	1*b*	Dry, heated	None	40	76.10	1.65	.....	....	......	....
	2 (total)	Dry.........	AC	100	76.46	1.26	.....	....	......	....
	2, lot 1	Dry.........	AC	50	77.18	1.48	.....	....	......	....
	2, lot 2	Dry.........	AC	50	75.74	2.03	.....	....	......	....
P. jeffreyi 26	2*a*	Dry.........	None	40	62.08	1.24	.....	....	......	....
	2*b*	Dry, heated	None	40	76.10	1.23	.....	....	......	....
	3*a*	Dry.........	None	40	62.90	1.95	.....	....	73.85	2.07
	3*b*	Dry, heated	None	40	74.30	1.50	.....	....	103.02	2.95
	1*a*	Dry.........	AC	40	76.48	2.53	.....	....	......	....
	1*b*	Wet.........	AC	40	75.65	2.24	64.10	1.45	111.58	2.72
	1*c*	Wet, heated	AC	40	76.78	2.35	67.93	2.03	109.18	2.97
	2	Wet.........	AC	40	74.15	2.10	.....	....	109.25	2.12
	3*aa*	Wet.........	AC	40	77.22	2.37	.....	....	112.83	3.11
	3*b*	Wet.........	60 days } In acidic	50	80.06	2.07	70.88	2.11	118.94	3.14
	3*c*	Wet.........	90 days } storage	50	81.56	2.31	.....	....	......	....
P. jeffreyi 10	1*a*	Dry.........	None	100	71.96	0.91	44.55	0.86	117.98	2.35
	2*a*	Dry.........	AC	100	85.88	1.96	75.74	1.52	117.98	2.35
	2*b*	Wet.........	AC	100	87.62	1.44	73.73	1.18	121.79	2.26
	2N	Wet.........	AC	100	85.31	1.65	69.80	1.21	113.81	1.86

6. Acidic storage causes some increase in pollen sizes after a period of time, which in the test was 60 days and 90 days. The differences between the first and the others are significant, but the difference between the 60 and 90 days is not (compare slide 3*aa* with slides 3*b*, *c* of *P. jeffreyi* 26 in appendix table 8*a*).

7. Fresh pollen grains, desiccated or not, show no significant difference in size after acetolysis (appendix table 8*b:* compare *P. muricata* 11, slides N*a*, *b* with slide N*c*).

8. Variations of size within the same species are not significant so long as the same treatment is given the samples.

APPENDIX TABLE 8*b*

VARIATION IN SIZE OF *Pinus muricata* POLLEN WITH DIFFERENT TREATMENT AND MOUNTING PROCEDURES

Species and sample number	Slide number	Treatment (all wet mounts)	N	X_1 (in μ)	2SE (in μ)	X_2 (in μ)	2SE (in μ)	X_3 (in μ)	2SE (in μ)
P. muricata 11	N*a*........	Desiccated at 300 °F in oven, then AC	100	69.98	1.32	66.40	1.25	99.80	1.45
	N*b*........		100	69.02	1.28				
	N*c*........	AC when moist	100	70.70	1.20				
P. muricata 12	*a*..........	AC when moist	30	61.70	1.31				
	b..........	AC when moist	30	64.00	1.75				
P. muricata 13	N*a*........	AC when dry	30	68.30					
	N*b*........	AC when moist	30	68.10					
	N*c*........	AC when moist	30	68.60					
	N (total)[a].	AC when moist	100	68.60	1.53	59.60	1.26	96.00	3.44

[a] N (total) = $a + b + c + 10$.

APPENDIX VI

Discussion on Papers by B. B. Christensen (1946) and S. T. Andersen (1960) Concerning Size of Pollen Grains

Christensen (1946) tested modern and fossil pollen grains of angiosperms, mostly *Corylus avellena* from various localities and from *postglacial* deposits, under desiccated and moist conditions, and under various treatments and combinations of treatments. His findings were as follows (the page and plate numbers refer to text of Christensen's report):

1. Whether or not the pollen grains were desiccated has no effect on the change of size from subsequent treatments (p. 17).
2. Pollen grains change size under different treatments (pl. 1).
3. Acetolysis expands the pollen grains.
4. The increase in size is instantaneous when the pollen grains contact the acetolyzed solution.
5. Heating of acetolyzed sample in the water bath induces a slight reduction in size (pp. 14–15).
6. *HF* treatment causes shrinkage in pollen size (p. 18).
7. Modern desiccated pollen grains treated by *KOH* show a reduction in size.
8. *KOH* treatment of fossil pollen grains causes an increase, though slight, in pollen size.
9. Fossil pollen grains of *Corylus avellena,* not treated, are different in size from recent ones. The difference is believed to be the result of fossilization, not taxonomic.
10. Acetolysis after *KOH* treatment has these effects: (*a*) It causes expansion of recent pollen grains to sizes much larger than acetolysis alone. (*b*) *KOH* treatment brings the recent pollen samples into a state approaching that of the fossil samples. "The increase of size [due to acetolysis] was just as great for the former as for the latter [p. 16]."
11. Acetolysis after *HF* reexpands the pollen grains to beyond its untreated size (Christensen, 1946, table 1: from 22.4μ to 21.07μ, then to 27.8μ).
12. The size of pollen grains from clayey gyttja after *HF* treatment is smaller than that from other types of deposits. This may be owing to original difference in size of fossil grains (pp. 17–18).

The results of my experiments in general confirm the findings of Christensen (see appendixes VII, VIII, below). For this study, I am immediately concerned with the possibility of comparing fossil samples treated by *HF* and acetolysis with modern samples treated by acetolysis only. Christensen's data for both modern and fossil samples are tabulated in appendix tables 9 and 10. The modern desiccated pollen samples gave sizes varying from 27.3μ to 29.2μ. The average is 28.043μ. The live and moist modern pollen samples and the one preserved in acetic acid gave slightly higher mean values of 29.2μ, 28.6μ, and 28.9μ, respectively. The overall mean of all modern samples is 28.30μ. The fossil sample treated by *HF* and acetolysis gave a mean of 27.80μ.

Christensen gives no variance of these mean values. It is not desirable to discuss their differences without the values of deviations. I, however, attempted to estimate the standard deviation by referring to the values given by Andersen (1960) for *Corylus avellena* pollen grains. The standard deviation for mean above 30μ is about 1.70μ, and that for sizes between 23μ and 26μ is 1.59μ. Assuming these values apply equally well to Christensen's data, I made a t test, using a small SD of 1.60μ, and found that there is no significant difference between the mean of the

APPENDIX TABLE 9

Average Mean Values of Pollen Size of *Corylus avellena* Samples from Christensen's Data (In μ)

Source of sample	Number of samples	Desiccated: No treatment	Desiccated: KOH	Desiccated: AC	Desiccated: KOH + AC	Alive and moist: No treatment	Alive and moist: KOH	Alive and moist: AC	In acetic acid: No treatment	In acetic acid: AC
Modern									(one sample only)	
København and Frederiksberg	2	26.50	25.65	*28.70*	31.45	34.05	25.90	28.90	25.20	28.90
All samples (see appendix table 10)...........	2 to 7	25.68	25.68	*28.04*	31.45	34.05	25.90	28.90	25.20	28.90
Fossil All samples....	3 to 5	22.70	23.20	33.87	34.72 or 34.38	.....	.....	.....		
Samples from alga gyttja only.........	2 to 3	22.37	22.83	33.87	34.80	.....	.....	.....		
Alga gyttja III only.........	1	22.40	.....	33.70	.....	.....	.....	.....	HF + AC only: 27.80	

Note: Difference between values in italics is not significant. Andersen's data for modern *Corylus avellena* treated by KOH + AC averages 32.18μ ± 1.71μ, not significantly different from modern 31.45μ in the table.

fossil treated by $HF + AC$ and the mean of modern samples that were only acetolyzed. If I had used the larger SD (1.70μ), the results would be even more favorable.

The computation of the t test is done as follows:

Assume $SD = 1.60\mu$ and $N = 100$; then $SE = 0.16\mu$.

1. For 6 degrees of freedom ($df = 6$, because there are seven samples in the desiccated group), the desirable confidence interval with t at 5 per cent level of significance ($t_{0.5}$) is

$$(2.447)\,(0.16) = 0.392\mu.$$

The mean for the desiccated group $= 28.043\mu$ (appendix table 10), and the confidence limits are from 27.648μ to 28.432μ.

2. For $df = 9$ (all ten modern samples), the interval is

$$(2.262)\,(0.16) = 0.362\mu.$$

The mean of the entire group of modern samples $= 28.30\mu$, and the confidence limits are from 27.938μ to 28.662μ.

3. For fossil sample, the mean value is $27.80 \pm 0.16\mu$, and the limits are from 27.48μ to 28.12μ.

All three confidence limits overlapped one another. There is therefore no significant difference between the means compared.

APPENDIX TABLE 10

PART OF CHRISTENSEN'S DATA ON MEAN SIZE OF *Corylus avellena* POLLEN GRAINS (In μ)

Kind of sample	Sample description	AC	HF + AC
	Desiccated		
Modern	1. København	29.2	....
	2. Frederiksberg	28.2	....
	3. Farum (*a*)	27.3	....
	4. Farum (*b*)	27.7	....
	5. Jylland	28.1	....
	6. Cluj, Rumania	28.4	....
	7. Kasan, Russia	27.4	....
	Alive and moist		
	8. København	29.2	....
	9. Frederiksberg	28.6	....
	In acetic acid		
	10. København	28.9	....
Fossil	From alga gyttja III	....	27.8
Modern	Average of desiccated samples	28.043	....
Modern	Average of all samples	28.300	....

Andersen investigated the relative merits of silicone oil and many other materials as mounting media. Data of measurements of modern and fossil *Corylus avellena* pollen grains treated by *KOH* and acetolysis were available for an evaluation.

Silicone oil: The mean value of *Corylus avellena* pollen grains under silicone oil (Andersen's table 3), treated by *KOH* + acetolysis, stored for 5 days, is $26.64\mu \pm 1.34\mu$. No sample size is given, but since Andersen used 100 in most of the cases, the sample size is taken as 100. If so, the confidence interval indicated by 2 times the standard error is 0.264μ. From Andersen's table 5, a modern sample from the same locality (Hareskoven), with the same treatment as the preceding example, gives a mean of $26.26\mu \pm 1.35\mu$, the confidence interval is then 0.270μ for a sample size of 100. Since this value was labeled an average of three, the sample size may be 300, in which case the same interval is only 0.156μ. These two samples are not significantly different.

Now take the modern sample from Draved, South Jutland, in Andersen's table 5. The mean is $25.30\mu \pm 1.14\mu$; the confidence interval is 0.228μ. Compare it with that the fossil sample of noncalcareous gyttja (appendix table 13, sample 54.1748) is significantly different only from the modern sample from Hareskoven and not from the modern sample from Jutland. The fossil sample of marl (appendix table 13, sample ϕ1561) is different from all modern samples. The modern sample from Jutland is significantly different from those from Seeland (appendix table 13).

APPENDIX TABLE 12

SEQUENTIAL TEST OF MEANS OF TEN MODERN AND FOSSIL SAMPLES OF *Corylus avellena* IN SILICONE OIL AFTER POTASSIUM HYDROXIDE + ACETOLYSIS TREATMENT

Sample	*X*	F4	F3	M6(J)	F2	M5(S)	M4(S)	F1	M3(av)	M2(S)
M 1 (S)	26.65	+	+	+	+	1.01 (0.89)	0.77 (0.83)	0.77 (0.78)	0.39 (0.72)	0.01 (0.6)
M 2 (S)	26.64	+	+	+	1.12 (0.89)	1.00 (0.83)	0.76 (0.78)	0.76 (0.72)	0.38 (0.6)	
M 3 (S, av)	26.26	+	+	0.96 (0.89)	0.74 (0.83)	0.62 (0.78)	0.34 (0.72)	0.34 (0.6)		
F 1	25.88	+	1.29 (0.89)	0.58 (0.83)	0.36 (0.78)	0.24 (0.72)				
M 4 (S)	25.88	+	1.29 (0.83)	0.58 (0.78)	0.36 (0.72)					
M 5 (S)	25.64	+	1.05 (0.78)	0.34 (0.72)						
F 2	25.52	+	0.93 (0.72)	0.22 (0.6)						
M 6 (J)	25.30	1.82 (0.72)	0.71 (0.6)							
F 3	24.59	1.11 (0.6)								
F 4	23.48									

NOTE: If a figure is larger than the one below it in parentheses, the two samples are significantly different. If the upper figure is smaller, there is no difference. Read row by row. If any difference (upper figure) is less than its *D* (figure below it), then read no farther either to the right of that difference along the row or below it in the column. Heavy zigzag line separates significance (+) from nonsignificance.

SYMBOLS: M, modern; S, Hareskoven; av, average; F, fossil; J, Jutland

The *F* ratio (appendix table 13, lower part) between (1) and (2) is 1.124, which is not significant, whereas the *F* ratios both between (1) and (3) and between (2) and (3) are significant. In order to show a significant difference between the means of modern and fossil, it is necessary to show a difference significantly greater than that between means of (1) and (2). In order to show a significant difference between fossil samples from different soils, it would be necessary to show a significant difference between (2) and (3). Hence Goyette's analysis confirmed that there is no significant difference between the means of modern and fossil samples and that the fossil means are significantly different because of soil differences. The latter part of the conclusion about soil differences confirms Andersen's findings (Andersen, 1960).

Glycerine and glycerine jelly: Andersen gives the data in appendix tables 14 and 15 on both modern and fossil pollen grain sizes of *Corylus avellena.* With one exception, all samples are treated by *KOH* and acetolysis. The exception, fossil 572, was treated by *HF* before acetolysis, and gave the smallest mean.

APPENDIX TABLE 13

COMPARISON BETWEEN FOSSIL AND MODERN SAMPLES OF POLLEN OF *Corylus avellena* FROM ANDERSEN'S DATA
(Pollen samples treated by potassium hydroxide and acetolysis)

Sample	Student's t value ($t_{.025} = 1.96$)					
	Fossil sample from Andersen				Modern sample	
	54.1629	56.551	54.1748	ϕ 1.561	Jutland	Hareskoven
Modern from Jutland	2.08	0.79	2.12	8.79	....	4.33
Modern from Hareskoven, Seeland	1.20	2.34	4.48	11.38	4.33	....

Analysis of variance

Source of variation	Degree of freedom	Sum of squares	Mean square
(1) Between like groups	4	391	98.75
(2) Between different groups	1	111	111.00
(3) Within groups	594	1327	2.23

$F_{(1,4)} = 1.124$; $F_{(4,594)} = 44.28$; $F_{(1,594)} = 49.78$.

Sequential test of mean shows that fossil 1175 is significantly different from the rest, and that fossil 572 ($KOH + HF + AC$) is not different from modern samples in Andersen's table 3 ($KOH + AC$). Fossil H4092 is not different from modern samples in his table 8 (appendix table 15, sample codes 3, 4, 5). It is therefore possible to identify two of the fossil samples by their mean values.

The fact that a fossil treated by $KOH + HF + AC$ is not different in size from a modern treated by $KOH + AC$ indicates that the interposed HF treatment is immaterial to final size of the pollen grains. This finding by Andersen also contradicts Christensen's finding (1946).

APPENDIX TABLE 14

DATA AND COMPUTATIONS FOR SEQUENTIAL TEST APPLIED TO SIZE OF *Corylus avellena* IN GLYCERINE AND IN GLYCERINE JELLY.

Source in Andersen, 1960	Sample	N	X_1 (in μ)	SD (in μ)	Treatment	Confidence Interval	SD	SD^2
Table 1, p. 8	Modern 1163	100	31.87	1.51	KOH + AC	0.302	1.51	2.2801
Quoted, p. 8	Christensen	100	31.50	....	KOH + AC	.....	....	......
Table 3	Modern, glycerine	100	30.14	1.85	KOH + AC	0.370	1.85	3.4225
Table 8, p. 24	Modern, water, glycerine	100	33.02	1.74	KOH + AC	0.348	1.74	3.0276
Table 8, p. 24	Modern, alcohol, glycerine	100	33.25	1.79	KOH + AC	0.358	1.79	3.2041
Table 8, p. 24	Modern, acetone, glycerine	100	33.30	1.85	KOH + AC	0.370	1.85	3.4225
Table 2, p. 8	Fossil 572	104	29.98	2.75	KOH + HF + AC	0.540	2.75	7.5625
Table 2, p. 8	Fossil 1175	100	34.93	1.87	KOH + AC	0.374	1.87	3.4969
Table 2, p. 8	Fossil H4092	107	33.47	3.17	KOH + AC	0.613	3.17	10.0489
							Total	36.4651
							Average	4.55813

$$Sx = \left(\frac{4.55813}{a(a-1)/2}\right)^{\frac{1}{2}}$$, where $a = 8$ (excluding second item in the above table, which has no SD value given), and $a(a-1)/2 = 28$

$$= (0.1628)^{\frac{1}{2}}$$
$$= 0.4035$$

S_x	Q	D
0.4035	3.86	1.56
0.4035	3.63	1.46
0.4035	3.32	1.34
0.4035	2.77	1.12

NOTE: See appendix table 11 for definitions of Q and D.

APPENDIX TABLE 15

SEQUENTIAL TEST OF ANDERSEN'S DATA OF *Corylus avellena* IN GLYCERINE AND IN GLYCERINE JELLY

Code	Sample	X (in μ)	Code 8	7	6	5	4	3	2
1	Fossil 1175	34.93	+	+	+	1.91 (1.56)	1.68 (1.46)	1.63 (1.34)	1.46 (1.12)
2	Fossil H4092	33.47	+	+	1.60 (1.56)	0.45 (1.46)	0.22 (1.34)	0.17 (1.12)	
3	Modern, acetone, glycerine	33.30	+	3.16 (1.56)	1.43 (1.46)	0.28 (1.34)	0.05 (1.12)		
4	Modern, alcohol, glycerine	33.25	+	+	1.38 (1.34)	0.23 (1.12)			
5	Modern, water, glycerine	33.02	+	+	1.15 (1.12)				
6	Modern 1163	31.87	1.89 (1.34)	1.73 (1.12)					
7	Modern, glycerine	30.14	0.16 (1.12)						
8	Fossil 572	29.98							

NOTE: The table is read as described in appendix table 12. Zigzag line separates sample means that are significantly different (+) from those of nonsignificance.

APPENDIX VIII

Fossil Pollen Size Changes under Various Treatments

To test the changes of pollen size, a postglacial fibrous peat sample A4-11 from Upper Springy Pond, Brewster, Maine, was used. The deposit is very rich in pollen, containing many grains of *Pinus* and *Betula,* smaller quantities of *Tsuga* and *Fagus,* and a few grains each of many other taxa.

Pollen count given by *KOH*- and acetolysis-treated sample showed roughly: *Pinus,* 44 per cent; *Betula,* 30 per cent; *Tsuga,* 9 per cent; *Quecus,* 3 per cent; *Abies* and *Larix,* 1 per cent; *Picea,* trace; and other arboreals, 8 per cent.

The pollen was first washed and strained out of the peat with water. It was then divided into six equal parts: 2 for *KOH,* 2 for *HF,* 1 for acetolysis, 1 for no treatment. Two samples, one each of the *KOH*- and *HF*-treated samples, were again subjected to acetolysis. Thus, a total of six different treatments were made.

The pollen grains from the nontreated group were first measured under a water mount, keeping the slide constantly wet during measuring. After completion, the cover glasses were carefully removed and melted glycerine jelly was dropped over the slides. The pollen grains were remeasured. No significant difference in size was found between the two different types of mounts.

The average of the pollen measurements from slides mounted by glycerine jelly and from the water mount was then taken to represent the original size of fossil pollen grains.

The pollen grains measured are *Pinus, Tsuga, Betula, and Fagus.* Only the breadths of corpus and of saccus were measured in the pine pollen grains. For all others the longest line across the grain was measured.

Appendix table 17 shows the mean size of pollen grains of each taxon under different treatment. Their comparison with one another is indicated by a proportion. A proportion smaller than 0.95 and one larger than 1.05 are limits of nonsignificance. These are equivalent to the confidence interval at the 5 per cent level.

Major conclusions from this study are as follows:

1. No significant difference is found between original size and sizes resulting from *KOH* treatment.
2. Acetolyzed and *KOH* + acetolyzed grains are significantly larger than the original and the *KOH*-treated grains.
3. *KOH* plus acetolysis gives the largest grain size. All types of grains so treated, except those of *Fagus,* are significantly different from grains treated only by acetolysis.
4. *HF* treatment diminishes the pollen sizes to a state not significantly different from the original. *Fagus* pollen size is the only *HF*-treated one that is significantly different from *KOH*-treated ones.
5. Acetolysis after *HF* gives (*a*) a pollen size significantly larger than one treated by *HF* only, and (*b*) significantly smaller than those treated by *KOH* and acetolysis, except for *Fagus,* but (*c*) shows no difference from those treated by acetolysis only.

The general conclusion confirms the findings of Series A5 fossil sample dis-

APPENDIX TABLE 17

Sizes of Different Pollen Grains from Fossil Peat A4-11, Upper Springy Pond, Brewster, Maine, under Various Chemical Treatments

(Asterisk indicates significance)

(All means and standard deviations in μ)

Fossil pollen genus	N	NO[a]		KOH		NO v. KOH (In proportion)	AC		NO v. AC	KOH v. AC	KOH + AC		NO v. KOH + AC	KOH v. KOH + AC	AC v. KOH + AC	HF		NO v. HF (In proportion)	HF + AC		HF + AC v. AC	HF + AC v. KOH + AC
		X	SD	X	SD		X	SD	(in proportion)		X_1 / X	SD	(in proportion)			X	SD		X	SD	(in proportion)	
Pinus																						
X_1	100	45.32	4.75	47.27	4.77	1.04	62.69	8.42	1.38	1.32	70.09	9.04	1.54	1.48	1.12*	45.56	5.16	1.00	60.44	6.97	1.04	1.16*
X_2	100	40.85	3.94	40.46	4.29	0.99	54.80	6.88	1.34*	1.35*	64.61	7.02	1.56	1.58	1.17*	41.15	3.92	1.01	55.43	5.16	0.99	1.15*
Tsuga	60	66.30	8.20	67.65	6.68	1.02	83.75	11.92	1.26*	1.23*	99.30	9.18	1.49	1.47	1.18*	68.15	6.45	1.03	83.85	8.08	0.99	1.18*
Betula	100	21.92	2.02	22.60	2.45	1.03	28.62	3.11	1.31*	1.26*	30.51	3.06	1.39	1.35	1.06*	22.21	2.84	1.01	27.82	7.44	1.03	1.09*
Fagus	60	29.15	2.00	29.60	2.45	1.01	36.45	3.61	1.24*	1.23*	36.68	3.70	1.26	1.24	1.00	28.13	2.40	0.96	35.17	2.92	1.03	1.04
Corylus avellena	...	22.6	....	22.80	....	1.01	34.6	....	1.53*	1.52*	35.8	....	1.55*	1.53*	1.01	.....	....	....	.....	....	....	....
(Christensen)	...	22.4	....	.....	....	....	33.7	....	1.50*	....	.....	....	....	....	....	21.0	....	0.93	27.80	....	1.21*	....

[a] NO = no treatment.

cussed in Appendix VII, but contradicts in a number of points the findings of Christensen (1946):

1. The acetolyzed and *KOH* + *AC* samples of fossil *Corylus avellena* are not significantly different, from one another, whereas in this series all except *Fagus* are significantly different.

2. The *HF* and acetolyzed samples of *Corylus avellena* are significantly different from the acetolyzed sample, whereas none of the pollen types in this series is different.

APPENDIX IX

The Density Separation of Pollen Grains

Dichlorodibromomethane (CBr_2Cl_2) has a specific gravity of 2.42. It is soluble in alcohol and can be diluted to any specific density desired. It is highly fluid and gives a clean separation without causing clumps.

Two samples of postglacial deposits from Upper Springy Pond, Brewster, Maine, were used for testing. Both samples came from a core, A4, which was from the northern end of the pond. Sample A4-5 was about 10 feet below the water level, and sample A4-13 was from the bottom some 24 feet below the lake surface.

Sample A4-13A was a clayey silt. Acetolyzed preparation yields many grains of *Picea,* some *Pinus,* and no pollen of angiosperms. The nontreated sample was crumbled into fine particles, and washed with absolute alcohol. Separation of pollen grains was achieved by three solutions of different specific gravities, 2.0, 1.7, and 1.4, respectively.

The separation by specific gravity of 2.0 yielded no pollen grains. The separation by specific gravity of 1.7 was good, giving many grains of *Picea* and *Pinus* but no pollen grains of angiosperms. The separation by specific gravity of 1.4 gave essentially the same picture with regard to *Picea* and *Pinus,* though in a better concentration, but it also yielded some grains of *Betula* and *Alnus.*

Spurred by this find, I decided to investigate the density of pollen grains, and found tentatively that the densities of saccate acetolyzed pollens grains are higher than those of many angiosperms. The densities of pollen grains of *Abies, Picea,* and *Pinus* are between 1.40 and 1.70, while those of angiosperms are between 1.40 and 1.50 (table 6).

Sample A5, a fibrous peat, had already been studied from an acetolyzed sample, and its pollen content was known from a count of 300 grains. It would be of interest to compare the results with those obtained through density separation of the same acetolyzed sample.

A part of the acetolyzed sample A5 was divided into four portions, which were placed in four test tubes and dehydrated, and dichlorodibromomethane solutions at densities of 2.0, 1.8, 1.6, and 1.4 were added, one each, to the tubes.

After mixing and centrifuging, the liquid in each tube was poured into four other tubes, diluted to a density of approximately 1.15, centrifuged, and decanted.

The residues of each separation were mounted in glycerine jelly and a pollen count of 250 grains each was made. The results showed (appendix table 18) that separation by a dilution of higher density gives proportionately more pollen grains of gymosperms than angiosperms, whereas at a lower density the number of angiosperms began to increase; and that the number of nonarboreal pollen grains (NAP) seemed to be particularly abundant with the 1.8 density solution.

1. The maximum percentages of pollen types occur at different densities. The overall result seems to indicate that saccate grains tend to be more readily separated at higher densities than pollen grains of angiosperms. This is also true of the NAP pollen grains. A solution of specific gravity 2.0 gives the highest percentages to *Picea;* a 1.8 solution shows *Pinus* and NAP at a maximum; a 1.6 solu-

tion is most favorable for *Tsuga* and other unknowns, presumably *Rosaceae;* and a 1.4 solution is best for *Betula, Fagus,* and *Fraxinus.*

2. The relative frequencies of the sample separated by specific gravity 2.0 solution more closely resemble the results from the acetolyzed standard. The sample separated with specific gravity 1.8 solution is the least similar to the acetolyzed standard.

APPENDIX TABLE 18

POLLEN COUNTS OF INDIVIDUAL SAMPLES FROM SEPARATION BY DIFFERENT DENSITIES, THEIR AVERAGES, AND THE STANDARD COUNTS
(All samples acetolyzed)

Genus (arboreals only)	Standard, acetolyzed (N = 300)	Pollen count of samples made from separation by CCl_2Br_2 at various specific gravities (N = 250; count in percent of N)				Average count (in percent)
		2.0	1.8	1.6	1.4	
Abies	2.3	0.8	0.8	1.6	0.4	0.9
Picea	5.7	5.2	2.4	2.4	1.6	2.9
Pinus	25.6	28.0	28.4	24.4	25.6	26.6
Tsuga	14.6	18.8	29.6	32.4	12.4	23.3
Larix	0.3	0.0	1.2	0.4	0.0	0.4
Betula	30.0	28.4	21.6	24.0	30.4	26.1
Carpinus	1.0	1.2	0.0	0.0	0.4	0.4
Quercus	6.7	1.6	0.8	0.4	0.8	0.9
Fagus	10.0	10.0	12.0	7.6	19.6	12.3
Ulmus	0.0	0.0	1.2	0.4	0.4	0.5
Tilia	0.0	0.4	0.0	0.0	0.0	0.1
Alnus	0.7	0.8	0.8	0.0	1.2	0.7
Fraxinus	0.7	0.8	0.0	0.8	2.0	0.9
Salix	0.7	0.0	0.0	0.0	0.0	0.0
Acer	1.0	0.8	1.2	1.2	0.8	1.0
Castanea	0.0	0.0	0.0	0.0	0.4	0.1
Unknown	0.7	3.2	0.0	4.4	4.0	2.9
Nonarboreal pollen	1.0	10.0	18.4	4.4	8.4	10.3
ρ = Rank coefficient or correlation with standard		0.898	0.695	0.761	0.746	0.830

3. In view of the fact that in clayey samples, such as A4-13A, pollen are far more numerous in number and type when a solution of specific gravity 1.4 is used, and the fact that in sample A4-5 different counts may be obtained from using liquids of different densities, it is suggested (*a*) that the separation process be made as a series of separations with different densities, and (*b*) that the representative pollen count be the total sum of all individual counts based on samples of different densities. If an individual count is to be 250, and there are four steps in separation, then the total count should be 1,000 grains.

4. The setting of the "acetolyzed-only" sample as a standard is of course purely arbitrary.

SYMBOLS AND ABBREVIATIONS

AC	Acetolysis.
C	Coefficient of variance. Standard deviation divided by sample mean. Usually given as a percentage.
F	Variance ratio. Mean square of sample means divided by mean square of individuals.
F test	A test using F, from R. A. Fisher, so named by Snedecor (1959, p. 244).
Γ	Gamma index. Used to measure degree of similarity or resemblance between two individuals, samples, or groups.
HO	Hypothesis.
Max	The upper end of a pollen spectrum, designated in code.
Min	The lower end of a pollen spectrum, designated in code.
μ	Mu. 1/000 of a millimeter.
N	Sample size or lot size; number of observations, measurements, or counts that constitute a sample.
$N\overline{X}$	Product of sample or lot size and mean.
$N\overline{\overline{X}}$	Product of sample or lot size and average mean.
R_1	Mean of corpus breadth divided by mean of saccus breadth.
R_2	Mean of saccus breadth divided by the same (=1).
R_3	Mean of overall size divided by mean of saccus breadth.
R_4	Mean of overall size divided by mean of corpus breadth.
ρ	Rho. Spearman's rank coefficient of correlation.
SD	Standard deviation of sample mean.
SE	Standard error of sample mean ($= \mathrm{SD}/N^{1/2}$).
Σ	Summation sign.
t, t-ratio, or Student's-t	A ratio derived by dividing the difference of population mean and sample mean by standard error, from W. S. Gosset (see Snedecor 1959, p. 45).
$\overline{X}$	Sample mean.
$\overline{\overline{X}}$	Average mean, mean of means.
$\overline{X}_1$	Mean corpus breadth.
$\overline{X}_2$	Mean saccus breadth.
$\overline{X}_3$	Mean overall breadth.
x	Lot or sample size of unknown A.
y	Lot or sample size of unknown B.
z	Lot or sample size of unknown C.

GLOSSARY

Class	A subdivision of a spectrum (a frequency distribution) into which raw scores (pollen counts) are entered. A class has a given range (the class interval) and a class midpoint by which the class is identified.
Class-C index	A composite index made up of class number and coefficient of variance combined and expressed as a hyhenated number.
Class interval	The range of size variation within a class.
Class midpoint	The central value of a range of sizes within a class.
Class number	Number of classes in a spectrum (see Range).
Confidence interval	Interval within which the mean may be expected to vary without significance; an interval within which the population mean would be expected to fall.
Cos A index	An index derived from the law of cosines, $a^2 - b^2 + c^2 = 2bc \cos A$, where a is the ratio of mean of corpus breadth to mean of saccus breadth (or, expressed in symbols, $a = R_1$); b is the ratio of mean of saccus breadth to mean of saccus breadth, which equals 1 (or, expressed in symbols, $b = R_2$); and c is the ratio of mean of overall size to mean of saccus breadth (or, expressed in symbols, $c = R_3$). The natural cosine value is used as the index.
Frequency	The total score of pollen grains within a class, lot, or sample.
Group A	A group of pines whose pollen grains have verrucoid distal membrane.
Group B	A group of pines whose pollen grains have psilate (or glabrous) distal membrane.
Lot	A synonym for sample, used to designate samples not numbering 100 in size. Expressed as N.
Mode	Location of maximum score or frequency on a spectrum. In grouped data, the location is designated by the class midpoint, that is, mode at class 71 or simply mode 71. In this work a class with a score of frequency more than 3 per cent higher than the frequencies of its neighboring classes is called a mode (i.e., in a series of five classes having frequencies of 3, 6, 10, 8, 3—a total of 30 grains in all—the class with 10 grains is a mode because it is higher than each of its neighboring classes by more than 3 per cent. For a mode to fall at either end of a spectrum, there must be at least a 5 per cent difference between it and the adjacent class.

Proportion	Ratio of two values multiplied by 100, or ratio of two values where the smaller one is considered as 100. The difference between the two values, if more than 5, is usually significant in this study of pine pollen means. For example, two values 50 and 100 have a ratio of 2 and a proportion of 200, and their difference is 100 per cent, which is very significant.
Range	The two extreme limits of a spectrum, defined by the class midpoints. In this work, the upper end is the maximum (or max); the lower end is the minimum (or min). If a class at either end of a spectrum is separated from the other classes by two or more classes devoid of scores (frequency = 0), that class and all the classes devoid of scores are not counted when reckoning range or class number of spectrum. For example, spectra of both *P. jeffreyi* 2 and 5 (appendix table 2) have a class number of 14, not 19 or 20, and their ranges are 5–11 and 7–13, respectively, and not 5–14 and 7–16. Range is expressed by the class midpoint of the maximum- and minimum-end classes and by code as shown in appendix tables 1 and 2.
Ratio	A measure for comparison of two values, by letting the smaller value equal unity (=1).
Ratio index	Combination of two significant digits from R_1 and R_4 expressed as a four-digit number.
Sample	A synonym for lot. Designates a collection of observations, scores, or frequencies with a standard size of 100 either as a number or, in certain specified cases, as a percentage.
Sample mean	Sum of individual scores divided by total number of individuals (= arithmetic average). Expressed as X.
Sample number	The call number of a pollen sample. For example, *P. aristata* 6A is sample 6A of pollen collection of *P. aristata*. The number designates a distinct collection, and the letter next to it signifies a subdivision of the same collection.
Sample size	Designates all samples with a standard size of 100. Expressed as N.
Score	Individual pollen count per class, lot, or sample.
Spectrum	Frequencies or scores arranged in numerical sequence by class midpoint.

REFERENCES

ALRIDGE, J.

1842. Memoir to determine the use of pollen in natural classification. J. Bot., (London) vol. 1, pp. 575–601.

ANDERSEN, S. T.

1954. Late glacial pollen-diagram from southern Michigan, U.S.A. Danmarks Geol. Undersøg., ser. 2, no. 80, pp. 140–155.

1960. Silicone oil as a mounting medium for pollen grains. *Ibid.*, ser. 4, vol. 4, no. 1.

ANDERSON, E.

1954*a*. "Efficient and inefficient methods of measuring specific differences." *In* Statistics and Mathematics in Biology. Ames: Iowa State Univ. Press, Pp. 93–106.

1951*b*. Några data on pollen variationen och pollenfertiliteten hos gran och tall. Svensk. Papperstidn., vol. 57, pp. 242–255.

AXELROD, D. I., and W. S. TING

1960. Late Pliocene floras east of the Sierra Nevada. Univ. Calif. Publ. Geol. Sci., vol. 39, pp. 1–118.

1961. Early Pleistocene floras from the Chagoopa surface, southern Sierra Nevada. *Ibid.*, pp. 119–194.

AYTUG, B.

1960. Quelques mensurations des pollens de *Pinus silvestris* L. Pollen et Spores, vol. 2, pp. 305–309.

1963. Diagnose des pollens de *Pinus silvestris* et *Pinus uncinata* des Pyrénées. *Ibid.*, vol. 4, pp. 283–296.

BANERJEE, U. C., J. R. ROWLEY, and MARY L. ALESSIO

1965. Exine plasticity during pollen grain maturation. J. Palynology, vol. 1, pp. 70–89.

BENT, A. M., and H. E. WRIGHT, JR.

1963. Pollen analysis of surface materials and lake sediments from the Chuska Mountains, New Mexico. Bull. Geol. Soc. Amer., vol. 74, pp. 491–500.

BUELL, M.

1946*a*. Size-frequency study of fossil pine pollen compared with herbarium preserved pollen. Amer. J. Bot., vol. 33, pp. 510–516.

1946*b*. A size frequency study of *Pinus banksiana* pollen. J. Elisha Mitchell Sci. Soc., vol. 62, pp. 221–228.

1947. Mass dissemination of pine pollen. *Ibid.*, vol. 63, pp. 163–167.

CAIN, S. A.

1940. The identification of species in fossil pollen of *Pinus* by size frequency determinations. Amer. J. Bot., vol. 27, pp. 301–308.

CHRISTENSEN, B. B.

1946. Measurement as a means of identifying fossil pollen. Danmarks Geol. Undersøg., ser. 4, vol. 3, no. 2, pp. 1–20.

CRANWELL, L. M.

1940. Pollen grains of the New Zealand conifers. New Zeal. J. Sci. and Tech., vol. 22, pp. 1*b*–17*b*.

1961. Coniferous pollen types of the southern hemisphere. I. Aberration in *Acmopyle* and *Podocarpus Dacrydioides*. J. Arnold Arbor., vol. 42, pp. 416–423.

CRITCHFIELD, W. B.

1957. Geographic variation in *Pinus contorta*. Maria Moors Cabot publ., no. 3. Cambridge: Harvard Univ. Press.

CUSHING, E. J.

1961. Size increase in pollen grains mounted in thin slides. Pollen et Spores, vol. 3, pp. 265–275.

DEEVEY, E. S.

1939. Studies on Connecticut lake sediments. I. A post-glacial climatic chronology for southern New England. Amer. J. Sci., vol. 237, pp. 691–724.

1942. Studies on Connecticut lake sediments. III. The biostratonomy of Linsley Pond. *Ibid.*, vol. 240, pp. 233–264, 313–338.

DELCOURT, A., W. MULLENDERS, and P. IIERART

1959. La préparation des spores et des grains de pollen, actuels et fossiles. Natur. Belg., (Brussels) vol. 40, pp. 89–120.

DYAKOWSKI, J.

1958. On the possibility of determination of the pollen of some species of the genus *Abies*. Acta Biol. Cracov. Bot., vol. 1, pp. 57–68.

ERDTMAN, G.

1931. Pollen statistics: A new research method in palaeoecology. Science, vol. 73, pp. 399–401.

1943. An introduction to pollen analysis. Waltham, Mass.

1956. "LO-analysis" and "Welcker's rule," a centenary. Svensk. Bot. Tidskr., vol. 5, pp. 135–141.

1957. Pollen and spore morphology: Plant taxonomy. Stockholm: Almqvist and Wiksell.

1960. The acetolysis method. Svensk. Bot. Tidskr., vol. 54, pp. 561–564.

1965. Pollen and spore morphology: Plant taxonomy. Gymnospermae, Bryophyta (Text). Stockholm: Almqvist and Wiksell.

ERDTMAN, G., and J. R. PRAGLOWSKI

1959. Six notes of pollen morphology and pollen-morphological techniques. Botanicka Notiser, vol. 112, pp. 175–184.

ERDTMAN, G., J. PRAGLOWSKI, and S. NILSSON

1963. An introduction to a Scandinavian pollen flora. Stockholm: Almqvist and Wiksell. Vol. II, pp. 1–90.

FAEGRI, K., and P. DEUSE

1960. Size variations in pollen grains with different treatments. Pollen et Spores, vol. 2, pp. 293–298.

FAEGRI, K., and J. IVERSON

1964. Textbook of pollen analysis. 2d rev. ed. New York: Hafner, 237 pp.

FEDEROVA, R. V.

1951. Influence des méthodes de préparation sur la taille des grains de pollen des différentes espèces de *Pinus*. Trav. Inst. Geog. Akad. Nauk. SSSR, vol. 50, pp. 137–141.

1952*a*. Lois quantitatives de la dispersion du pollen dans les roches superficielles des voies flûviales. *Ibid.*, vol. 52, pp. 91–103.

1952*b*. Dispersion des spores et pollens par les cours d'eaux. *Ibid.*, pp. 46–72.

FERRE, Y. DE

1952. Les formes de jeunesse des *Abietacées* ontogenie-phylogenie. Trav. du Lab. Forest de Toulouse. T. 2, première section, vol. 3, art. 1, pp. 1–284.

GARRISON, W. L., and D. F. MARBLE

1961. The structure of transportation networks. Unpublished report for the U. S. Army Transportation Res. Command, by the Transportation Center at Northwestern University, p. 140. Cited by K. J. Kansky. Structure of transportation networks, Dept. Geogr. Res. Paper 84, Univ. of Chicago (1963).

GAUSSEN, H.

1960. Les Gymnospermes, actuelles et fossiles. Trav. du Lab. Forest de Toulouse. Fasc. 6, chap. 9, pp. 1–272.

GRAY, J., *et al.*

1965. Techniques in palynology. *In* F. Kummel and D. Raup, Handbook of paleontological techniques. San Francisco: W. H. Freeman and Co. Pp. 471–708.

GRAYSON, J. F.

1954. Evidence of four pine species from fossil pollen in Michigan. Ecology, vol. 35, pp. 327–331.

GREEN, P. S.

1955. Pollen grain size in *Nasturtium* and *Cakile*. Trans. Bot. Soc. Edin., vol. 34, pp. 289–304.

HALLER, J. H.
1962. Variation and hybridization in *Ponderosa* and *Jeffreyi* pines. Univ. Calif. Publ. Bot., vol. 34, no. 2, pp. 123–166.

HANSEN, H. P.
1947. Post-glacial forest succession, climate and chronology in the Pacific North-west. Trans. Amer. Philos. Soc., vol. 37, pp. 1–130.

HARRIS, W. F.
1957. Pollen of Nothofagus: Variations in size and number of aperture from flower to flower on the same tree. New Zeal. Sci. Tech., vol. 37, pp. 635–638.

HORMANN, H.
1929. Die pollenanalytische Unterscheidung von *Pinus montana, P. silvestris,* and *P. cembra.* Osterr. Bot. Z., vol. 78, pp. 215–228.

IVERSON, J., and J. TROELS-SMITH
1950. Pollen morfologiske definitioner og typer. Danmarks Geol. Undersøg. ser. 4, vol. 3, pp. 1–54.

JAESCHKE, J.
1935. Zur frage der Artdiagnose der *Pinus silvestris, P. montana,* und *P. cembra* durch Variation-statistische pollenmessungen. Beih. Bot. Centralblatt, vol. 52*b*, pp. 622–633.

JENNRICH, R. I., and P. F. SAMPSON
1964. Stepwise discriminant analysis. Biomedical Computer Programs, BMDX67. Health Sciences Computing Facility, Department of Preventive Medicine and Public Health, School of Medicine, Univ. of California, Los Angeles.

KREMP, G. O. W.
1965. Morphologie encyclopedia of palynology. Tucson: University of Arizona Press. Pp. 1–186.

MARTENS, P., and L. WATERKEYN
1962. Structure du pollen (âile) chez des conifers. La Cellule, vol. 62, pp. 173–222.

MARTIN, A. R. H.
1959. South African palynological studies. I. Statistical and morphological variations in the pollen of the South African species of *Podocarpus.* Grana Palynologica, vol. 2, pp. 40–68.

OSTLE, B.
1963. Statistics in research. 2d ed. Ames: Iowa State Univ. Press. 585 pp.

POLLOCK, J. B.
1906. Variation in the pollen grains of *Picea excelsa.* Amer. Nat., vol. 40, pp. 159–170.

PONS, A.
1956. Application de la méthode des diagrammes de dispersion symbolique à d'analyse pollinique. Naturalia Monspeliensia, Ser. Bot., vol. 8, pp. 177–188.
1964. Contribution palynologique a l'étude de la flore et de la végétation Pliocènes de la région Rhodanienne. Thèses présentées a la Faculté des Sciences de l'Université de Montpellier. Paris: Masson.

ROBERTSON, A.
1904. Spore formation in *Torreya californica.* New Phytol., vol. 3, pp. 133–140.

RUDOLPH, K.
1935. Mikrofloristische Untersuchung Tertiärer Ablagerungen im Nordlichen Bohmen. Beih. Bot. Centralblatt., vol. 54, pp. 1–244.

SNEDECOR, G. W.
1959. Statistical methods. 5th ed., Ames: Iowa State Univ. Press. 534 pp.

SRIVATAVA, S. K.
1961. Morphology of normal and some abnormal pollen grains of *Pinus roxborghii* Sarg. Grana Palynologica, vol. 3, pp. 130–132.

STARK, P.
1927. Die moore des Badischen Bodenseegebiets. II. Beih. Naturforsch. Ges. Freiburg, vol. 28, pp. 1–238.

SZAFER, W.

1949. Studies on the genus *Tsuga* Carr. in the Tertiary of Europe. Bull. Acad. Polon. Sci. et Lettres, Ser. B, vol. 1, pp. 23–51.

TAKEOKA, M.

1960. On the classification between *Tsuga sieboldii* and *Tsuga canadensis*, based on the fine surface structure of pollen membrane. Sci. Repts. Kyoto Pref. Univ. Agr., vol. 12, pp. 83–86.

TING, W. S.

1965. The saccate pollen grains of *Pinaceae* mainly of California. Grana Palynologica, vol. 6, pp. 270–289.

TING, W. S., and C. C. TSENG

1965. Fine structure of pollen wall of *Pinus balfouriana* Grev. et Balf. Pollen et Spores, vol. 7, pp. 9–10.

UENO, J.

1957. Relationships of Genus *Tsuga* from pollen morphology. J. Inst. Polytech., Osaka City Univ., Ser. D., vol. 1, pp. 191–196.

1958. Some palynological observations of *Pinaceae. Ibid.*, Ser. D, vol. 9, pp. 163–277.

1960. Studies of pollen grains of gymnospermae. *Ibid.*, vol. 11, pp. 109–136.

VAN CAMPO-DUPLAN, M.

1950. Recherches sur la phylogenée des Abietinées d'après leurs grains de pollen. Trav. du Lab. Forest de Toulouse, vol. 4, art. 1, pp. 1–177.

WENNER, C. C.

1953. Investigation into the possibilities of distinguishing the pollen of the various species of *Betula* in fossil material. Stockholm Hogskolar Geol. Inst., no. 103, ed. 75, pp. 367–380.

WHITEHEAD, D. R.

1964. Fossil pine pollen and full-Glacial vegetation in southeastern North Carolina. Ecology, vol. 45, pp. 767–776.

WILSON, I. T., and R. H. KOSANKE

1940. The mocrofossils in pre-Kansan peat deposit near Belle Plain, Iowa. Torreya, vol. 40, pp. 1–5.

WILSON, L. R.

1965. Tetralalogical forms in pollen of *Pinus flexilis* James. J. Palynology, vol. 1, pp. 106–110.

WODEHOUSE, R. P.

1935. Pollen grains. Reprinted 1960. New York and London: McGraw-Hill. 574 pp.

YAMAZAKI, T., and M. TAKEOKA

Electromicroscope investigation on the surface of the pollen membrane, based on the carbon replica method.

1957. J. Jap. Forest Soc., vol. 39, pp. 427–434.

1958*a*. *Ibid.*, vol. 40, pp. 7–11.

1958*b*. *Ibid.*, vol. 40, pp. 154–159.

1958*c*. Sci. Repts. Saikyo Univ. Agr., vol. 10, pp. 28–32.

1959*a*. J. Jap. Forest Soc., vol. 41, pp. 125–129.

1959*b*. Sci. Repts. Kyoto Pref. Univ. Agr., vol. 11, pp. 86–90.

1962. Electron microscope investigations of the fine details of the pollen grain surface in Japanese gymnosperms. Grana Palynologica, vol. 3, pp. 3–12.

PLATES

PLATE 1

Pollen Sculpture
(In phase contrast)

Fig. 1. Verrucose distal membrane of *Pinus lambertiana* 2 (× 920).

Fig. 2. Psilate distal membrane of *Pinus contorta* 7 (× 920).

Fig. 3. W pattern on cappa of *Pinus muricata* 1 at high focus (× 1,800).

Fig. 4. Y pattern on cappa of *Pinus ponderosa* 2 at high focus (× 1,800).

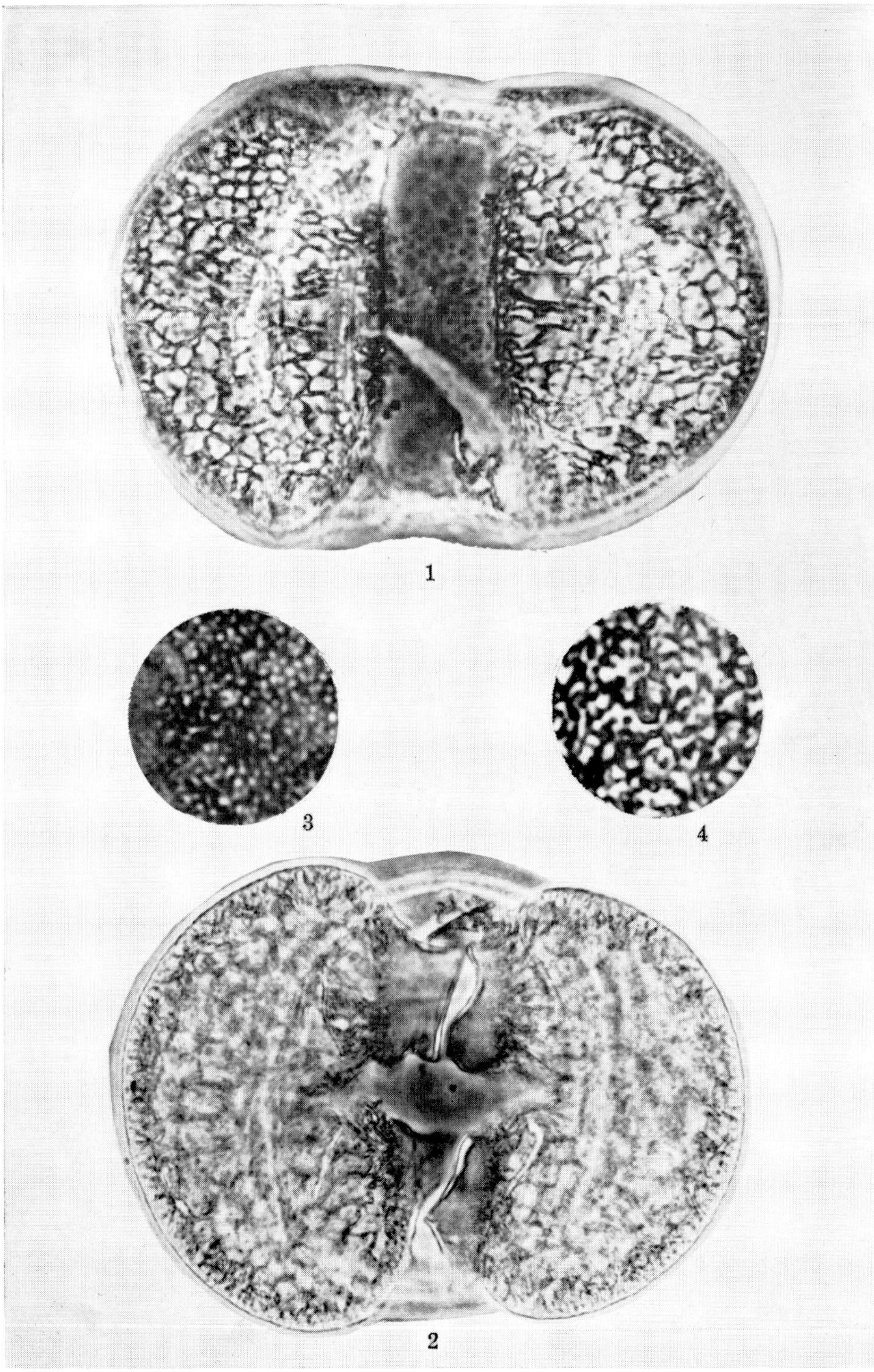
1
3
4
2

PLATE 2

Pinus Pollen with W Pattern
(In phase contrast)

Fig. 1. *Pinus albicaulis* 14 (× 920).

Fig. 2. *Pinus flexilis* 3 (× 920).

Fig. 3. *Pinus lambertiana* 2 (× 920).

1
2
3

PLATE 3

Pinus Pollen with W Pattern
(In phase contrast)

Fig. 1. *Pinus monticola* 1 (× 920).

Fig. 2. *Pinus monophylla* 2. Note distal membrane is verrucose (× 920). (Aberrant ?)

Fig. 3. *Pinus lambertiana* 3 (× 920).

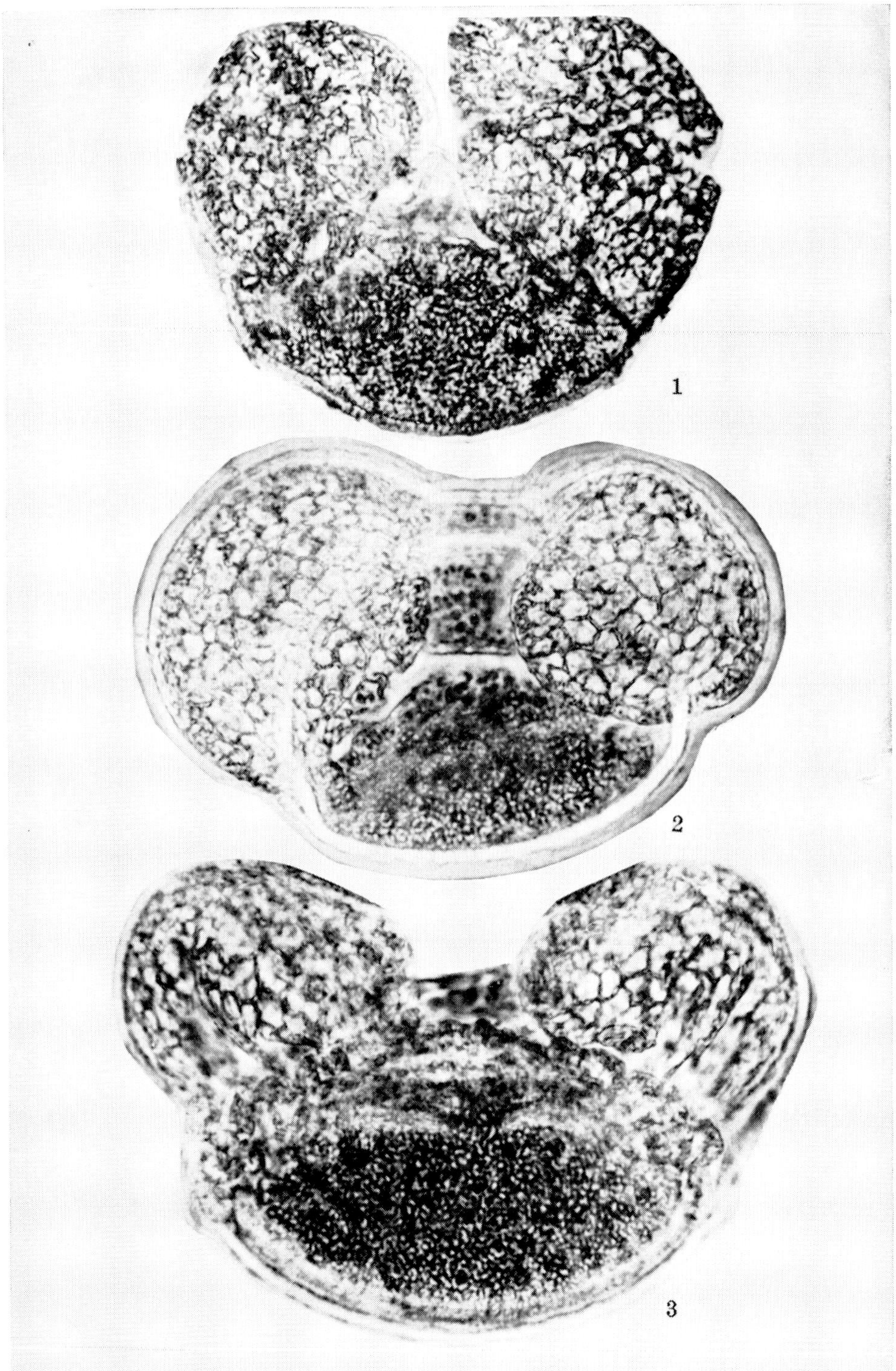

PLATE 4

Pinus Pollen with W Pattern
(In phase contrast)

Fig. 1. *Pinus radiata* 2 (× 920).

Fig. 2. *Pinus attenuata* 4 (× 920).

Fig. 3. *Pinus sabiniana* 2. Note distal membrane is glabrous (× 920).

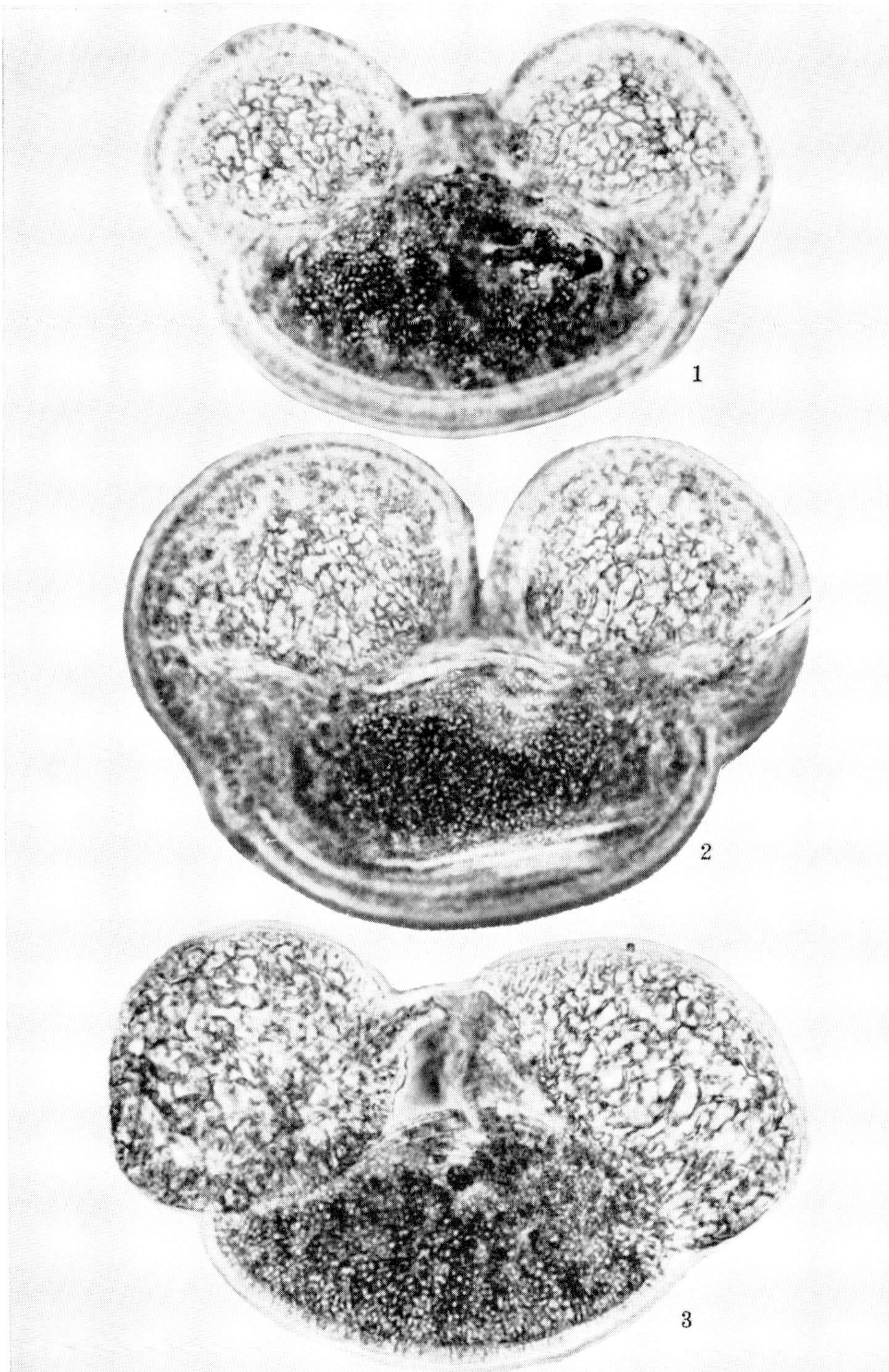
1
2
3

PLATE 5

Pinus Pollen with W and Y Patterns
(In phase contrast)

Fig. 1. *Pinus remorata* 3. W pattern (× 920).

Fig. 2 *Pinus ponderosa* 6 (× 920).

Fig. 3. *Pinus jeffreyi* 2. Note all distal membranes are glabrous (× 920).

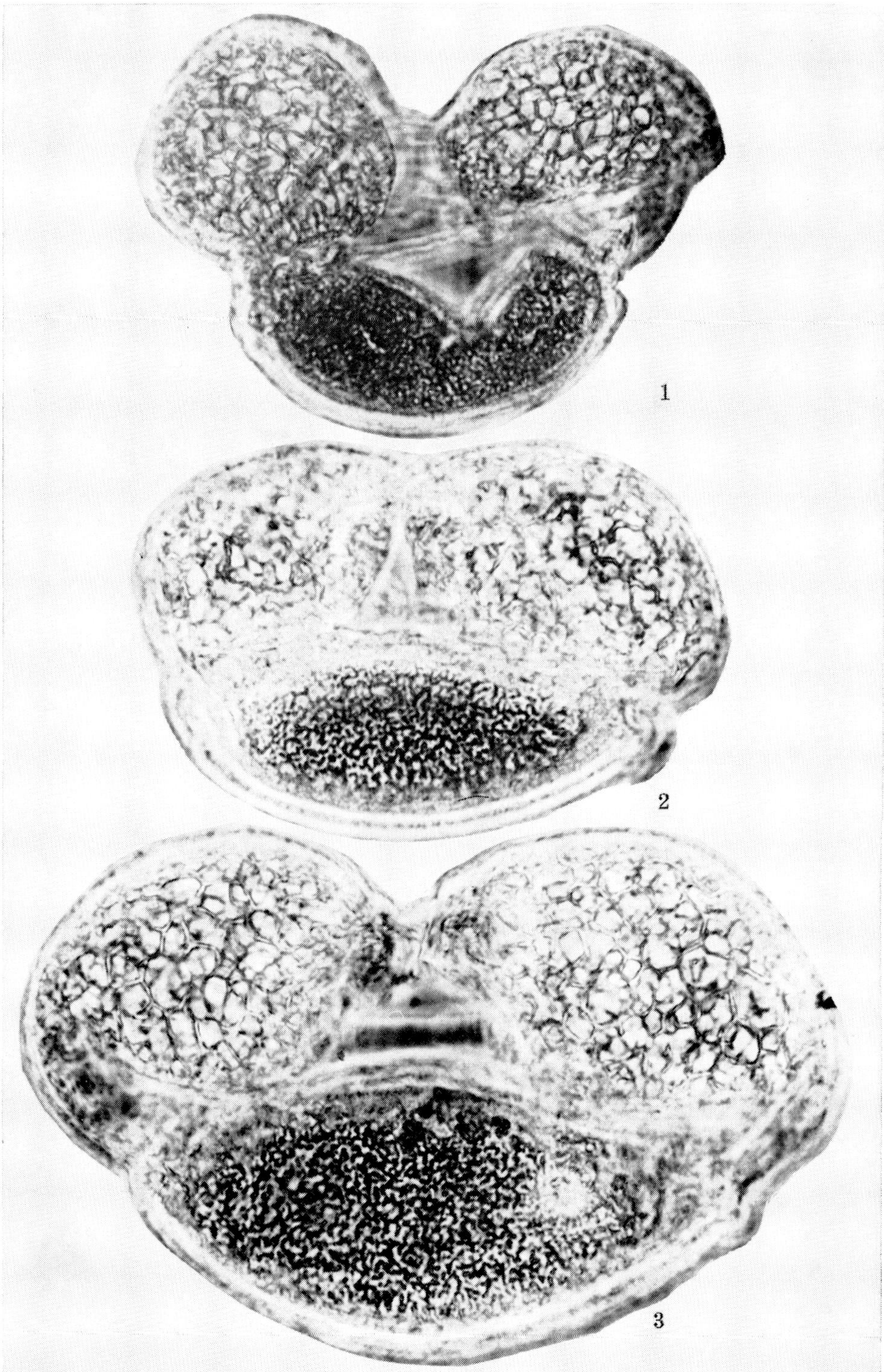

1

2

3

PLATE 6

Pinus Pollen with Y Pattern
(In phase contrast)

Fig. 1. *Pinus balfouriana* 4. The shape is very typical (× 920).

Fig. 2. *Pinus aristata* 6. Exine undetermined, probably Y pattern but inset (*a*) appears to be W pattern. Note distal membrane is verrucose (× 920).

Fig. 3. *Pinus contorta* var. *bolanderi* 1. Note distal membrane is glabrous (× 920).

Fig. 4. *Pinus murrayana* 5. Note distal membrane is glabrous (× 920).

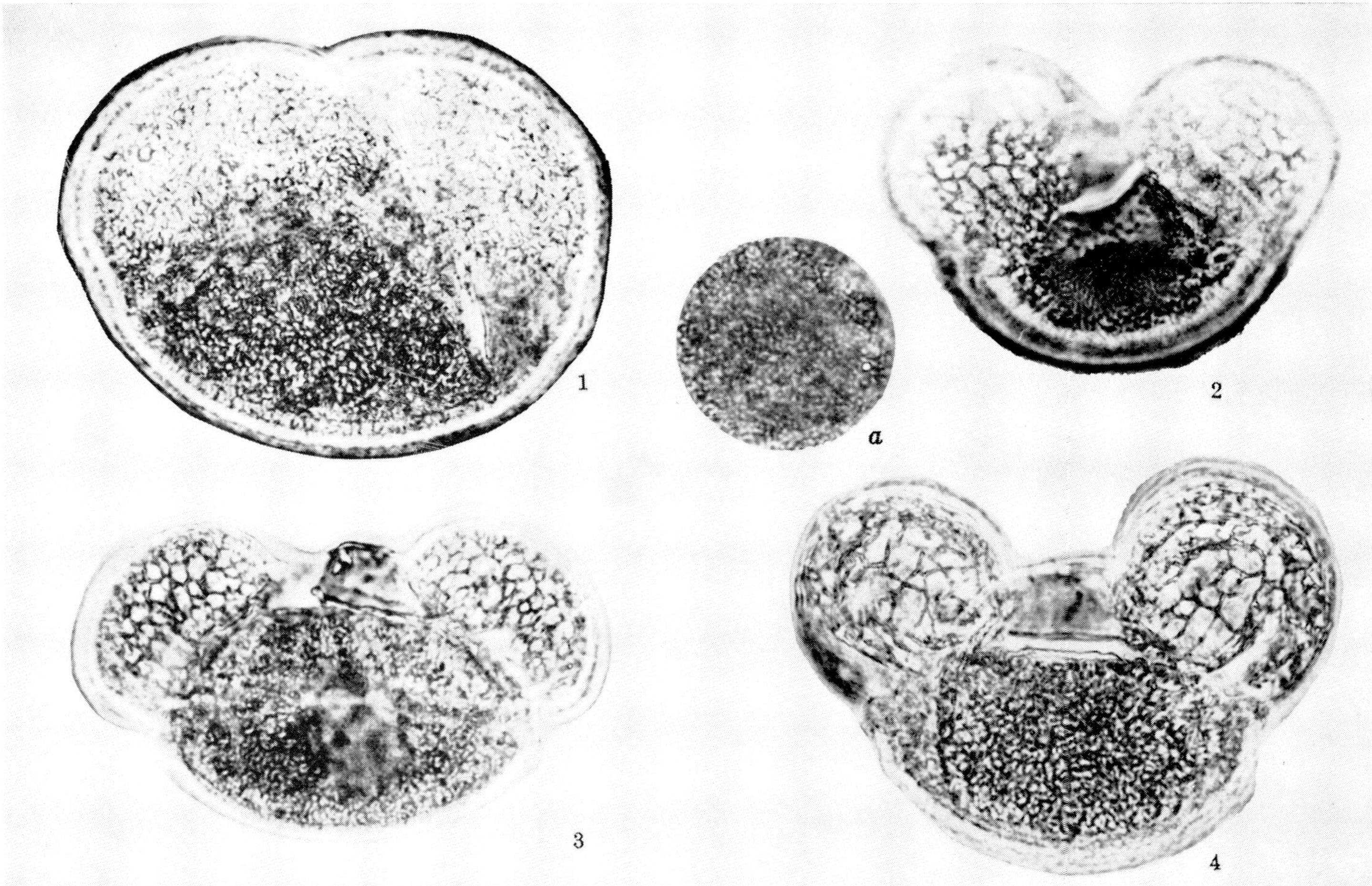
1
a
2
3
4

PLATE 7

Abnormal Forms and *Pinus* Pollen with W Pattern
(In phase contrast)

Fig. 1. Trisaccate *Pinus jeffreyi* 3, showing tear on cappa (triradiate mark ?) (×920).

Fig. 2. *Pinus lambertiana* 2, showing unisaccate grain (×575).

Fig. 3. *Pinus muricata* 1 (×920).

Fig. 4. *Pinus torreyana* 4 (×920).

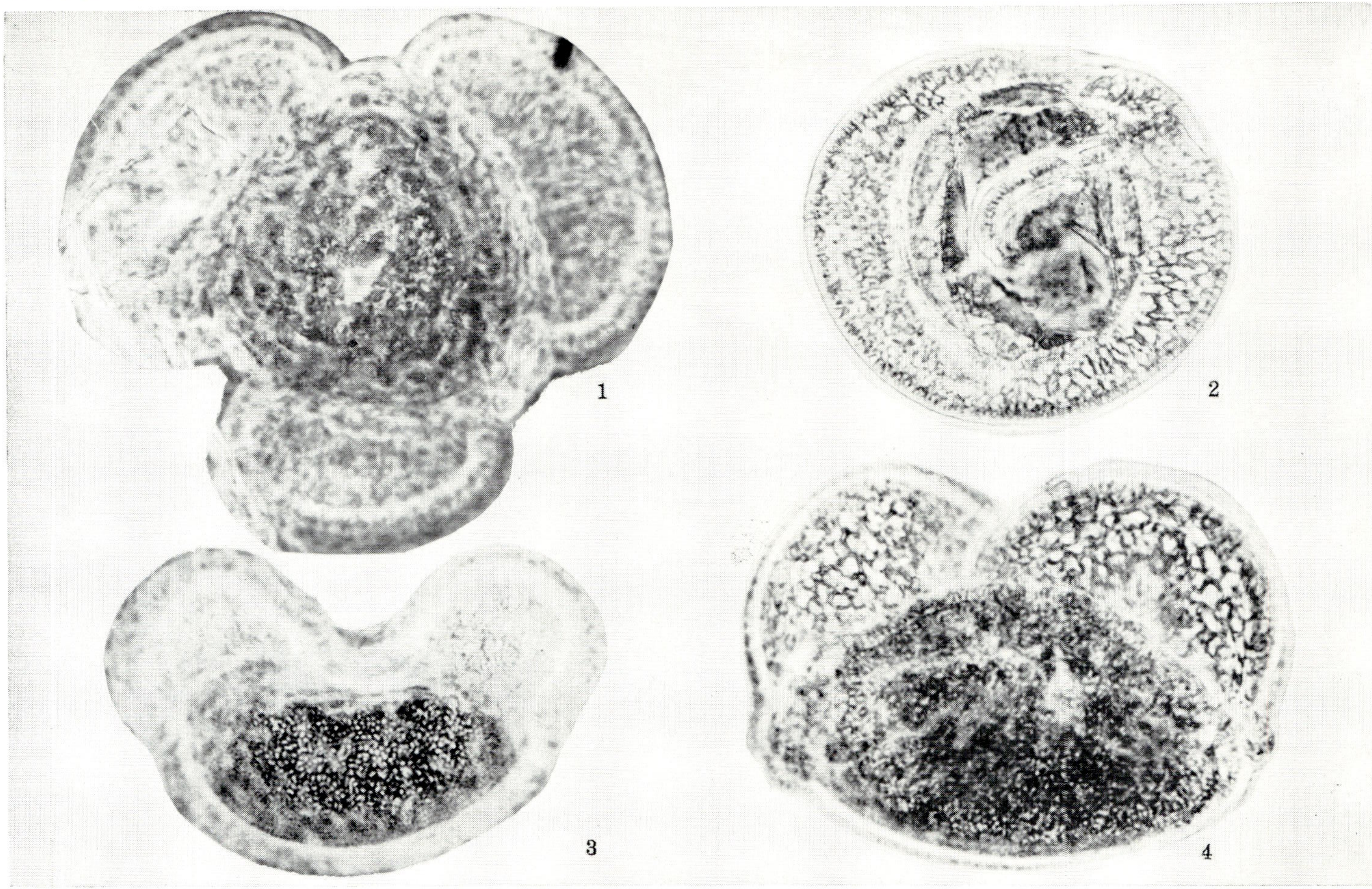
1
2
3
4